Contaminant Transport Through Aquitards: A State-of-the-Science Review

Awwa Research Foundation

About the Awwa Research Foundation

The Awwa Research Foundation (AwwaRF) is a member-supported, international, nonprofit organization that sponsors research to enable water utilities, public health agencies, and other professionals to provide safe and affordable drinking water to consumers.

The Foundation's mission is to advance the science of water to improve the quality of life. To achieve this mission, the Foundation sponsors studies on all aspects of drinking water, including supply and resources, treatment, monitoring and analysis, distribution, management, and health effects. Funding for research is provided primarily by subscription payments from approximately 1,000 utilities, consulting firms, and manufacturers in North America and abroad. Additional funding comes from collaborative partnerships with other national and international organizations, allowing for resources to be leveraged, expertise to be shared, and broad-based knowledge to be developed and disseminated. Government funding serves as a third source of research dollars.

From its headquarters in Denver, Colorado, the Foundation's staff directs and supports the efforts of more than 800 volunteers who serve on the board of trustees and various committees. These volunteers represent many facets of the water industry, and contribute their expertise to select and monitor research studies that benefit the entire drinking water community.

The results of research are disseminated through a number of channels, including reports, the Web site, conferences, and periodicals.

For subscribers, the Foundation serves as a cooperative program in which water suppliers unite to pool their resources. By applying Foundation research findings, these water suppliers can save substantial costs and stay on the leading edge of drinking water science and technology. Since its inception, AwwaRF has supplied the water community with more than $300 million in applied research.

More information about the Foundation and how to become a subscriber is available on the Web at **www.awwarf.org.**

Contaminant Transport Through Aquitards: A State-of-the-Science Review

Prepared by:
J.A. Cherry and **B.L. Parker**
University of Waterloo, Waterloo, Ontario, Canada

K.R. Bradbury, **T.T. Eaton**, **M.B. Gotkowitz**, and **D.J. Hart**
Wisconsin Geological and Natural History Survey, Madison, WI

and

M.A. Borchardt
Marshfield Clinic Research Foundation, Marshfield, WI

Sponsored by:
Awwa Research Foundation
6666 West Quincy Avenue, Denver, CO 80235-3098

Published by:

Awwa Research Foundation

American Water Works Association

IWA Publishing

DISCLAIMER

This study was funded by the Awwa Research Foundation (AwwaRF). AwwaRF assumes no responsibility for the content of the research study reported in this publication or for the opinions or statements of fact expressed in the report. The mention of trade names for commercial products does not represent or imply the approval or endorsement of AwwaRF. This report is presented solely for informational purposes.

CONTENTS

TABLES

FIGURES

FOREWORD

The Awwa Research Foundation is a nonprofit corporation that is dedicated to the implementation of a research effort to help utilities respond to regulatory requirements and traditional high-priority concerns of the industry. The research agenda is developed through a process of consultation with subscribers and drinking water professionals. Under the umbrella of Strategic Research Plan, the Research Advisory Council prioritizes the suggested projects based upon current and future needs, applicability, and past work; the recommendations are forwarded to the Board of Trustees for final selection. The foundation also sponsors research projects through the unsolicited proposal process; the Collaborative Research, Research Applications, and Tailored Collaboration programs; and various joint research efforts with organizations such as the U.S. Environmental Protection Agency, the U.S. Bureau of Reclamation, and the Association of California Water Agencies.

This publication is a result of one of these sponsored studies, and it is hoped that its findings will be applied in communities throughout the world. The following report serves not only as a means of communicating the results of the water industry's centralized research program but also as a tool to enlist the further support of the nonmember utilities and individuals.

Projects are managed closely from their inception to the final report by the foundation's staff and large cadre of volunteers who willingly contribute their time and expertise. The foundation serves a planning and management function and awards contracts to other institutions such as water utilities, universities, and engineering films. The funding for this research effort comes primarily from the Subscription Program, through which water utilities subscribe to the research program and make an annual payment proportionate to the volume of water they deliver and consultants and manufacturers subscribe based on their annual billings. The program offers a cost-effective and fair method for funding research in the public interest.

A broad spectrum of water supply issues is addressed by the foundation's research agenda: resources, treatment and operations, distribution and storage, water quality and analysis, toxicology, economics, and management. The ultimate purpose of the coordinated effort is to assist water suppliers to provide the highest possible quality of water economically and reliably. The true benefits are realized when the results are implemented at the utility level. The foundation's trustees are pleased to offer this publication as a contribution toward that end.

Walter J. Bishop
Chair, Board of Trustees
Awwa Research Foundation

Robert C. Renner, P.E.
Executive Director
Awwa Research Foundation

ACKNOWLEDGMENTS

The authors of this report are indebted to the following water utilities and individuals for their cooperation with this project:

Madison Water Utility, Madison, WI., David Denig-Chakrof
Waukesha Water Utility, Waukesha, WI.
Madison Metropolitan Sewerage District, Madison, WI., David Taylor

Dr. R. A Freeze contributed invaluable insight, discussions, and reviews during the project.

EXECUTIVE SUMMARY

Aquifers and aquitards make up groundwater flow systems, with aquitards having much lower permeability than aquifers. Aquitards help protect underlying aquifers, and water-supply wells completed in them, from contamination. The degree of this protection depends on many factors, including contaminant type, hydrogeologic setting, flow system dynamics, and local effects of groundwater pumping. *Aquitard integrity* refers to the capability of an aquitard to provide protection to an underlying aquifer. Determination of aquitard integrity is an important aspect of groundwater management because without it, predictions of future groundwater quality cannot be reliable.

RESEARCH OBJECTIVES

The Awwa Research Foundation (AwwaRF) funded this study to determine how water utilities and their technical consultants could evaluate the susceptibility of aquifers to contamination and how aquitards help protect water supply wells and aquifers from contamination. This work was carried out by a team of investigators from the Wisconsin Geological and Natural History Survey, the University of Waterloo, and Marshfield Medical Research Foundation. The key objectives of the study included the following:

1. Conduct a review of all relevant literature on aquitards, including hard-to-find works from unpublished theses, project reports, and European sources.

2. Using this literature, produce a "state-of-the-science" report of aquitard evaluation relevant for the purposes of understanding and predicting chemical and microbial transport to underlying aquifers (this document).

3. Evaluate and demonstrate selected investigative techniques for aquitards at existing sites in Ontario and Wisconsin where background information and regional groundwater models already exist.

4. Develop a methodology and technical guidance document for practical, cost-effective aquitard evaluation and risk analysis (a companion document; Bradbury and others, 2006).

The research team conducted an extensive literature review of published and unpublished sources and also consulted professional colleagues from across North America and Europe. The literature review suggested several key field sampling and instrumentation techniques for aquitard investigation, and the team carried out field studies to evaluate the utility and cost-effectiveness of these techniques. Most of this work occurred at field sites near Madison, Wisconsin, where municipal water supply wells penetrate a regional bedrock aquitard. Work at these field sites generated significant new field data and important experience with investigative techniques. Based on the literature review and field studies, the team developed two reports that together meet AwwaRF's objectives of understanding how aquitards affect contamination susceptibility. This report, *Contaminant Transport Through Aquitards: A State-of-the-Science Review* summarizes the current knowledge of aquitard science with emphasis on aspects most relevant to groundwater resources use and management. This report serves as the technical foundation for a second report *Contaminant Transport through Aquitards: Technical Guidance for Aquitard Assessment*. This second report is intended to provide specific guidance concerning investigations of aquitard integrity, and summarizes specific technical methodologies, categories of data collection and synthesis, and examples that will be of use in evaluating aquitards. Accordingly, the second report is arranged as a technical manual rather than as a research report.

CONCLUSIONS

Current understanding of aquitards

1. The traditional hydrogeologic and engineering literature usually presents an oversimplified treatment of aquitards. Textbooks depict aquitards as thick, uniform units of considerable horizontal extent, but most field studies demonstrate aquitards typically have large spatial variability in thickness and lithology. They also commonly contain secondary features such as fractures, macropores, or other heterogeneities that strongly influence contaminant migration.

2. Aquitards that have more than 10 to 15 percent by weight clay-sized particles behave hydrogeologically as clayey units (e.g., extremely low matrix hydraulic conductivity that does not increase appreciably with larger percentage of clay-size particles). Clayey aquitards that have no preferential pathways, such as fractures, root holes or other discontinuities for groundwater flow and contaminant migration, and that are laterally extensive, provide the greatest degree of protection to underlying aquifers because the small intergranular permeability severely limits fluid movement even when hydraulic gradients are large. In these aquitards the groundwater age, estimated by groundwater velocity calculations and isotopic age indicators, is typically thousands of years or older. Migration of dissolved contaminants is typically governed by molecular diffusion driven by contaminant concentration gradients and can occur over time scales of hundreds or thousands of years. Field studies indicate that such diffusion-controlled aquitards exist in some regions of North America.

3. Preferential pathways greatly diminish the integrity of many aquitards because the velocity of contaminant migration along these pathways can be many orders of magnitude greater than in the matrix material. Vertical fractures are the most common type of preferential pathway, and even fractures with typically small apertures (<100 μm) can be important contaminant pathways. Although the integrity of an aquitard generally increases with increasing thickness, even large aquitard thickness (>50 m) should not be used without other lines of evidence to conclude that there is absence of open, fully penetrating fractures. The geologic origin of the aquitard and the post-depositional history are more important indicators of aquitard integrity than the thickness.

4. The most useful information about an aquitard is usually internal data, such as hydraulic head measurements or water samples collected from piezometers situated inside the aquitard. External measurements, such as hydraulic head measured in aquifers above and below an aquitard, provide only indirect evidence of aquitard performance. Many

commonly used methods for characterizing the hydraulics of aquitards (for example, leaky-aquifer pumping test analyses) are of limited value in predicting contaminant movement through aquitards, because the inherent simplifying assumptions are often inappropriately applied in the analyses of contaminant movement. These external methods rarely identify the types or location of migration pathways in aquitards.

5. The hydraulic properties of aquitards are generally very poorly known. Direct field measurements, at appropriate scales, of important hydraulic parameters such as vertical hydraulic conductivity and effective porosity are usually necessary to assess aquitard effectiveness and predict contaminant movement.

6. Of the various types of contaminants that threaten aquifers, Dense NonAqueous Phase Liquids (DNAPLs) have the greatest propensity to move through fractured aquitards. Particulate contaminants very small in size, such as viruses, theoretically have the next largest propensity to travel quickly through fractured aquitards because they are not retarded by matrix diffusion. Dissolved contaminants have the least propensity to travel through fractured aquitards because their migration is retarded by matrix diffusion and other attenuating processes, such as sorption, precipitation and/or degradation.

Vulnerability of water supply wells

1. Aquitards are critical to protecting water supply wells from contamination. In general any well constructed with no aquitard between a contamination source and the well screen (or open borehole) is at great risk.

2. With respect to the protection that aquitards offer for water supply wells finished beneath them, assessments of the probability of contamination will be appropriate, meaningful, and useful only if the likely contaminant pathways through the aquitard are identified and characterized. Unfortunately, identifying these pathways in the field can be extremely difficult and expensive. However, recently developed innovative investigative sampling and monitoring tools (e.g., borehole imaging, flowmeter logging, depth-discrete multilevel monitoring systems, detailed core sampling for contaminants and/or isotopes) can provide essential information for aquitard assessment at reasonable cost. Assessment of the aquitard integrity and the vulnerability of underlying aquifers will improve as use of these innovative monitoring systems becomes more common.

3. The extremely small size of viruses (<100 nm) compared to the probable sizes of fractures and macropores observed in aquitards (5-50 μm) suggests aquitards may not always provide barriers to virus transport. The few studies done to date have documented enteric viruses present in groundwater at depths up to 47 m. At this time, predicting whether a particular aquitard will prevent adjacent aquifers from becoming virus contaminated is not possible. The most confident assessment and assurance of sanitary quality would come from sampling a confined aquifer for pathogenic enteric viruses, preferably monthly throughout the year. Research is needed to address gastrointestinal virus transport and survival through aquitards.

Opportunities for improved aquitard assessment

Appropriate hydrogeologic and geophysical tools and methods for assessing aquitards have been developed and are now available for systematic application in standard practice. However, at present only a few aquitards have been investigated using many of these tools. Progress has been hindered by the lack of textbooks and guidance documents focused on aquitards. The guidance document (Bradbury and others, 2006) prepared as a companion to this report is intended to fill part of this literature deficiency.

The aquitard literature focuses mostly on unlithified clayey aquitards (e.g., clayey Quaternary deposits); however, aquitards composed of shale and other types of rock (lithified aquitards) also provide protection to bedrock aquifers in many regions. The greater abundance of publications concerning unlithified aquitards is largely due to the much lower cost and greater ease of study of these shallow aquitards. Because the published literature is focused on aquitards of glacial or glaciolacustrine origin, this report has a similar bias. There is need for more research studies directed at bedrock aquitards and aquitards in non-glaciated regions.

RECOMMENDATIONS

1. Hydrogeologists should designate geologic units or zones as aquitards based on hydraulic data rather than stratigraphy or lithology. Aquitards are often mistakenly defined and depicted in textbooks and reports as being composed of entire geologic formations or stratigraphic units when this is rarely the case. Often only a few key beds in a much thicker stratigraphic unit provide the restrictions in flow and/or contaminant migration to warrant designation as an 'aquitard'. As a consequence, aquitards can be thinner, and adjacent aquifers thicker, than stratigraphic considerations alone would indicate.

2. Hydrogeologic and engineering studies conducted when designing water supply wells should strive to collect sufficient data on adjacent aquitards to allow a reasonable assessment of vulnerability to contamination and prevent well designs that breach aquitards and cross connect aquifers.

3. The most important field data needed in assessment of the effectiveness of an aquitard is a series of measurements of hydraulic head at different depths inside the aquitard. These measurements should include both depth-discrete and temporal data.

4. Aquitard investigations should include the use of environmental tracers such as chloride, oxygen-18, and deuterium. The distribution of these hydrochemical and isotopic constituents of the groundwater both inside and external to aquitards can be diagnostic of aquitard conditions for solute transport.

5. Recently developed innovative investigative sampling and monitoring tools (e.g., borehole imaging, flowmeter logging, depth-discrete multilevel monitoring systems, incremental core sampling) can provide essential information for aquitard assessment at reasonable cost. Some of the tools suitable for indurated aquitards are not suitable for non-indurated aquitards.

CHAPTER 1

INTRODUCTION

BACKGROUND AND DEFINITIONS

Aquitards are perhaps the most important yet most poorly understood components of groundwater flow systems. Aquitards control recharge and contaminant transport to adjacent aquifers, but methods for assessing their physical properties are not well developed. Although many public water supplies draw water from confined aquifers thought to be well protected by overlying aquitards, efforts to verify aquitard integrity in the context of groundwater protection are rare. The role of aquitards in groundwater protection depends on the interrelationship of geology, groundwater flow, and contaminant type.

Aquitards are geologic deposits of sufficiently low hydraulic conductivity, and sufficient areal extent, thickness and geometry, to impede groundwater flow between or to aquifers (Figure 1.1). Aquitards, unlike aquifers, do not generally supply economic quantities of water to wells but commonly determine flowpaths and serve as storage units for groundwater contributed to aquifers (Bates and Jackson 1987). Although aquitards are often called *confining units* in the literature and in textbooks, the terms *aquitard* and *confining unit* are not synonymous. Aquitards can hydraulically confine aquifers beneath them, in which case the hydraulic head in the underlying aquifer is above the elevation of the base of the aquitard. However, not all aquitards are confining and a large confining head does not imply strong aquitard integrity.

This report is based on a review of the scientific literature about aquitards with the goal of describing, in the context of aquitard integrity, aquitard properties and the nature of groundwater flow and contaminant migration influenced by aquitards. The report takes a hydrogeologic perspective in which the groundwater domain encompasses both aquitards and aquifers as the components of a single system in which the two components are interdependent and interactive in the context of flow and contaminant migration. Major water supplies are obtained from aquifers and consequently hydrogeologic investigations most commonly focus on flow and contaminant transport in aquifers. However, the low-permeability geologic formations that constitute aquitards are also important components of groundwater flow systems. Materials that make up aquitards volumetrically constitute the bulk of sedimentary geologic deposits (Potter, Maynard, and Pryor 1980). Aquitards are usually thought to provide underlying aquifers with protection from contamination.

Textbooks and monographs often provide only minimal information about aquitard integrity in the context of aquifer protection, however the broader literature about aquitards is voluminous. Results of aquitard studies are documented in specialized journals, conference proceedings and unpublished reports, but have not been synthesized in monographs or other documents intending comprehensive review. Much of the information in the aquitard literature derives from studies not directed specifically at assessment of aquitard capacity for protection of underlying aquifers. The scientific and engineering literature about aquitards is voluminous because aquitards have relevance to several important topics of societal relevance such as water resources, geotechnical engineering, and subsurface waste disposal. Aquitards have received attention from hydrogeologists because they govern the sustainable yield of many aquifers and it is this aspect that was the initial impetus decades ago for hydrogeologic studies of aquitards. Hydrogeologic studies of contaminant migration in aquitards began more recently and this topic has received much less attention than contaminant migration in aquifers. Surficial clayey aquitards have been studied intensively by geotechnical engineers to minimize harm to buildings and other structures caused by consolidation. A few aquitards have been investigated intensively for selection of low-permeability environments for waste disposal. For example, in Europe, countries such as, Belgium, France, Spain, and Switzerland are conducting intensive studies of deep clay and shale aquitards to select safe long-term repositories for radioactive waste.

This report summarizes the state of aquitard science focused on groundwater use and management with particular attention to the role of aquitards in protection of groundwater resources from contamination. Therefore, the literature review for this report made greater use of the hydrogeology literature than that of other disciplines. Chapter 1 of this report presents introductory information about aquitards, discusses their hydrogeologic functions, describes potential contaminants and warns of some common misconceptions. Chapter 2 discusses the role of aquitards in regional flow systems, and presents a geologic perspective on their hydrogeologic setting. Chapter 3 summarizes groundwater movement through aquitards. Chapter 4 describes contaminant transport and Chapter 5 discusses biological contaminants, with a focus on virus transport. Chapter 6 summarizes the current understanding of aquitards with emphasis on the protection of water supply wells from contamination.

HYDROGEOLOGIC FUNCTIONS OF AQUITARDS

Aquitards exert critical controls on groundwater flow systems because regional flow systems are hydraulically continuous (Tóth 1995). Aquitards have a wide range of three-dimensional geometries (i.e., thickness and extent) that determine groundwater flow and recharge patterns (Fortin, van der Kamp, and Cherry 1991; Macfarlane et al. 1994). Aquitards can restrict recharge from the land surface, and their hydraulic properties can cause very long response times to changes in groundwater flow, such as the response of water levels in wells to pumping (Alley et al. 2002, Husain et al. 1998). Regional-scale hydraulic conductivity in aquitards determines the distribution of steady-state hydraulic head in underlying confined aquifer systems (Belitz and Bredehoeft 1990). Groundwater can be a mixture of younger water flowing in aquifers and older water from aquitards, with average age increasing with relative aquitard volume (Bethke and Johnson 2002).

Aquitards have differing roles in the context of various environmental problems. For example, aquitards have been the focus of intensive study to assess their potential for long-term storage of industrial or radioactive waste (Boisson et al. 2001, Bonin 1998, Cherry 1983, Cherry et al. 1989, Flint et al. 2001, Gautschi 2001, Gillham and Cherry 1982, Mazurek et al. 1998), and many techniques of aquitard characterization were developed in these investigations. Fine-grained sedimentary rock aquitards have an important role as source beds for petroleum resources. These aquitards can determine the location of oil and gas fields in the subsurface by forming "traps" in regional flow systems (Hearn 1996).

One of the most important roles of aquitards in the groundwater system is the protection of aquifers from chemical or microbial contamination (Ponzini, Crosta, and Giudici 1989). For individual wells, concern over the level of protection provided often motivates the delineation of wellhead protection areas (WHPAs) to reduce the risk of contaminant transport from the land surface (U.S. Environmental Protection Agency 1987). However, the presence of aquitards can complicate the delineation of wellhead protection areas because of complexities they impose on the groundwater flow system (Martin and Frind 1998). Assessment of the risk of contaminant transport through aquitards should not be based entirely on wellhead protection considerations because of aspects of contaminant migration discussed later in this report. An aquitard can form the surficial geologic unit overlying an aquifer, or it can occur between a shallow aquifer and a deeper aquifer. In either case, the aquitard can provide greater protection from contamination to an underlying aquifer than that provided to aquifers without overlying aquitards (Domenico and Schwartz 1990, Fetter 2001).

Specialized Roles of Aquitards

Molecular diffusion commonly has a much more important influence on solute transport in aquitards than in aquifers where advection dominates (Gillham and Cherry 1982). The distribution of natural dissolved constituents and the isotopes of water (Farvolden and Cherry 1988; Hendry and Wassenaar 1999; Hendry, Wassenaar, and Kotzer 2000) and contaminants in aquitards (Parker, Cherry, and Chapman 2004) can result from long-term diffusion with little or no influence by groundwater flow. Aquitards subjected to long-term groundwater contamination can in turn become long-term contaminant sources to aquifers due to reverse diffusion instigated by aquifer remediation (Chapman and Parker N.d., Freeze and McWhorter 1997, Geistlinger et al. 1998, Liu 1999, Liu and Ball 1999, Liu and Ball 2002).

Aquitards can also serve as important biological and chemical interfaces to aquifers. They serve as sources for reactive minerals and exchangeable ions (Back 1985) as well as zones for biogeochemical reactions (McMahon 2001). Even in aquitards with very low permeability, there is evidence that natural bacteria may survive for geologic time periods (e.g., Lawrence et al. 2000). Aquifer geochemistry is in part dependent on microbial processes requiring organic compounds, which can originate from aquitards (McMahon and Chapelle 1991). Aquitards can provide information about the geologic history of their deposition and the deposition of associated geologic units in cases where the aquitard pore water contains isotopes or other indicators of geologic age (e.g., Hendry and Wassenaar 1999, Husain 1996). In some circumstances the hydraulic head distribution in or across buried aquitards can be governed by osmosis because the aquitard behaves as a semi-permeable membrane (Neuzil 2000; Cey, Barbour, and Hendry 2001). More specialized functions of aquitards are likely to be discovered as research focuses on this under-studied component of groundwater systems.

TYPES OF CONTAMINANTS

The three general types of contaminants in groundwater are:

1. Aqueous-phase (i.e., dissolved in water)
2. Light or dense non-aqueous phase liquid (LNAPL's, i.e., petroleum products and DNAPL's, i.e., chlorinated solvents, croesote)
3. Particulate (i.e., colloid size particles – inert, or biologically active)

Table 1.1 shows categories and examples within the categories. Nearly all research relevant to contaminant behavior in aquitards has focused on aqueous phase contaminants, but in recent years DNAPLs in aquitards have also received attention (e.g., Hinsby et al. 1996; Jørgensen et al. 1998; Parker, Gillham, and Cherry 1994; Parker, Cherry, and Chapman 2004). Considerable variability in contaminant mobility exists within a given contaminant type because of major differences in chemical properties. The chemical compounds in the NAPL type are also found in the aqueous category because all NAPLs are water soluble to some degree. Because many NAPLs have density and viscosity much different from groundwater, NAPL flows in ways aqueous contaminants cannot. DNAPL migration as an oily phase is an important aspect in the assessment of aquitard integrity, given the propensity for DNAPL to sink much deeper in the subsurface.

Table 1.1 Typical anthropogenic contaminants

General Types	Categories	Examples
Aqueous Phase (dissolved constituents)	Inorganic (primarily ionic)	
	Major ions	Cl^-, SO_4^{2-}, ClO_4^-
	Nutrients	NO_3^-, PO_4^-
	Trace elements	"heavy metals" As, Pb, Cd, Cr
	Organic (non-ionic + ionic)	PCE, TCE, BTEX, PCB's, pesticides, MTBE, and 1,4-dioxane, etc.
	Radioactive	^{3}H, ^{90}Sr, ^{137}Ce
NAPLs (immiscible liquids)	DNAPLs – Much more dense than water	Chlorinated solvents TCE, PCE, DCM, etc.
	DNAPLs – Slightly more dense than water	Liquid Hg, PCB's, some pesticides Creosote and coal tars
	LNAPL – Less dense than water	Petroleum products Fuel oils, jet fuels, petroleum
Particulates (colloids)	Biologically active	Bacteria, viruses
	Chemically active	Colloid (size) facilitated transport – "sorption"

Contaminants in the particulate type include viruses, bacteria and any other organisms that may migrate in the subsurface. The least is known about particulate contaminants even though pathogenic organisms such as some viruses species have potential to migrate large distances in the subsurface. Although the potential importance of particulate contaminants is large, research concerning their public health effects and occurrence in groundwater is still at an early stage.

Chapter 4 describes the different mechanisms of aqueous and non-aqueous phase contaminant transport through aquitards, while Chapter 5 gives the latest understanding of the presence and ability of biologically active particulates to migrate through aquitards.

Aqueous Contaminants

Dissolved contaminants are the most common type found in groundwater wells. Each dissolved chemical species can behave very differently depending on its specific chemical properties. For example, some dissolved contaminants such as Cl^- are entirely non-reactive in groundwater. Non-reactive dissolved contaminants are transported by groundwater without influence of any chemical, biological or radiological processes. As these contaminants migrate, their concentrations are influenced by flow (advection), diffusion and mechanical dispersion. Most dissolved contaminants are reactive to some degree and a common type of reaction is partitioning (i.e., sorption) between the aqueous and solid phases. Sorption refers to the influences of cation exchange, absorption, adsorption and co-precipitation. Sorption does not result in loss of total chemical mass from the hydrogeologic system but does cause the contaminant front to migrate more slowly than non-reactive contaminants and reduces aqueous phase concentrations in parts of the plume. For example, cationic metals such as lead commonly have diminished mobility due to sorption, as do most organic contaminants. Poly-chlorinated biphenyls (PCBs), which typically enter the subsurface as DNAPLs, are only sparingly soluble and dissolved PCBs from the DNAPL are so strongly sorbed they generally do not migrate appreciable distances away from the DNAPL zones.

Another aqueous contaminant category includes those contaminants influenced by natural degradation, either abiotic or biotic. Degradation processes diminish the total contaminant mass, which causes both attenuation of concentrations and mass along the migration path and retardation of the rate of contaminant front advancement. In some cases, a contaminant is degraded, but the degradation products are also contaminants. For example, trichloroethylene (TCE) may degrade to cis 1,1 dichloroethylene (cis-DCE) or vinyl chloride (VC), both of which may be more hazardous to human health than TCE. Hence, degradation is not always a beneficial process.

NAPLs

The second contaminant type, NAPLs, includes all liquids immiscible in water (i.e., oily liquids). NAPLs are only rarely found in water samples from wells, however they play an important role in groundwater contamination because they commonly migrate considerable distances in the subsurface, and then dissolve into the flowing groundwater resulting in contaminant plumes typically spread much further than the DNAPL. There are two NAPL categories: liquids less dense than water (LNAPLs), and liquids more dense than water (DNAPLs). Gasoline and fuel oil are examples of LNAPLs, and chlorinated solvents, creosote and PCB liquids are examples of DNAPLs. An LNAPL, such as a lubricant oil, is composed of many chemical compounds including chain hydrocarbons and benzenes, toluene and xylenes, and may also contain halogenated organic species, such as PCBs occurring as impurities co-dissolved in the petroleum product. The volume edited by Weyer (1992) contains chapters concerning various aspects of LNAPL and DNAPLs in the subsurface. Pankow and Cherry (1996) describe the nature and behavior of DNAPLs in groundwater, but provide only minimal information about DNAPLs in aquitards. Because NAPLs usually move under the primary influence of their density, they commonly migrate as immiscible liquids in directions different from the groundwater flow direction.

Although LNAPLs and DNAPLs are in the same general contaminant category (NAPLs), they exhibit very different subsurface behavior and there can even be large differences in behavior between NAPL sub-categories. For example, TCE and PCB liquids are both DNAPLs, but TCE dissolved from DNAPL is commonly found in groundwater while PCBs dissolved from DNAPL are not. Since dissolved TCE can be transported large distances by groundwater and dissolved PCBs remain near their source (due to sorption), TCE has a much greater potential to cause widespread groundwater contamination. Because of the buoyancy effect, LNAPLs have much less tendency than DNAPLs to migrate downward through aquitards. However, in some circumstances, continued release of LNAPL to the subsurface can cause sufficient build-up of LNAPL free-product thickness to cause LNAPL to be driven downward in fractures much below the water table. Oliveira and Sitar (1985) described a case where LNAPL was found far below the water table in a clayey aquitard in California.

Of the three general categories of contaminants, DNAPLs have the greatest propensity to migrate through fractured aquitards and cause impacts on underlying aquifers, because they can move downward in fractures under the influence of the driving force caused by their own density (Kueper and McWhorter 1991). Also, most DNAPLs are less viscous than water, which enhances their propensity for flow in the subsurface (Pankow and Cherry 1996). DNAPLs, particularly chlorinated solvents, can move downward through some aquitards even where groundwater flow is upward. For such DNAPL movement to occur, the aquitard must have preferential pathways allowing DNAPL flow. However, very strong upward gradients can prevent downward DNAPL movement (Chown, Kueper, and McWhorter 1997). Chapter 4 provides more detail about subsurface NAPL behavior and impacts in the context of aquitards.

Particulates

The particulate type includes contaminants exhibiting a wide variety of behaviors. Particulates include extremely small (colloidal) particles comprised of mineral and organic matter typically derived from plants, wood or coal, or other biologically active "particles" such as bacteria or viruses. Particles that are not biologically active are not usually contaminants on their own, however, they may carry and acclerate the migration of attached (sorbed) contaminants (i.e., facilitated transport). For example, a mobile colloid composed of mineral matter may carry sorbed hazardous metals such as cadmium. Particulate contaminants are transported in the subsurface by the bulk movement of water, however, some or all of the particles may be filtered out by obstructions along the flow paths (e.g., pore exclusion) or attachment by electrostatic forces to the geologic media along the flow paths. Thus, fewer particles are mobile with increasing distance from the source. Other mechanisms may also act to attenuate the number of particles migrating, or to alter their chemical or biological activity.

Upward groundwater flow in an aquitard would prevent particulate contaminants from moving downward, and therefore an aquitard with such flow would have good integrity with respect to this type of contaminant as long as the upward flow condition persists. Particulate contaminants are not influenced by molecular diffusion and therefore the attenuating effects of diffusion important for aqueous contaminants in aquitards do not apply. Field studies of particulate contaminant behavior are rare.

MEASURES OF AQUITARD INTEGRITY

Current interest in aquitards is focused on protection of underlying aquifers from contamination from surface and near-surface sources. *Aquitard integrity* refers to the degree to which an aquitard is protective of groundwater quality in underlying aquifers. Figure 1.2 illustrates different types of pathways for dissolved contaminant migration from a near-surface contaminant source to a well. The conceptual scenarios illustrated in Figure 1.2 are arranged from most vulnerable to contamination (Figure 1.2a) to least vulnerable (Figure 1.2e) (e.g., from poorest aquitard integrity to best integrity). In general, a water-supply well situated in an aquifer not overlain by an aquitard is relatively vulnerable to contamination (Figure 1.2a). The presence of an aquitard above the well screen generally decreases vulnerability, however the magnitude of the decrease depends strongly on the nature and extent of the aquitard. For example, Figure 1.2b shows an aquitard above the well screen, but the areal extent of the aquitard is too small for the aquitard to provide much protection. The groundwater pathway from the contaminant source to the well is diverted due to the aquitard but nevertheless the pathway is relatively short and fast. Figure 1.2c shows a laterally extensive aquitard, however a gap (i.e., geological window) in the aquitard allows easy groundwater passage to the well; the aquitard may cause a lengthening of the travel path but does not greatly diminish vulnerability of the well. The geologic material forming the window may be similar to the aquifer above or below, or it may be comprised of geologic media that would retard contaminant migration. Therefore, although a window is a negative feature in an aquitard, the aquitard nevertheless may provide considerable protection to the underlying aquifer. Figure 1.2d represents a scenario where the aquitard is laterally extensive but fully penetrating vertical fractures throughout allow rapid contaminant migration. However, whether or not contaminants travel slowly or quickly in the fractures depends on many factors, including fracture aperture and contaminant type. The presence of open, vertical fractures generally diminishes aquitard integrity, however, the degree of diminishment depends strongly on the contaminant type and the fracture characteristics. In Figure1.2e, the aquitard has no fractures, geologic windows or any other type of preferred extensive pathway for flow and therefore the aquitard prevents rapid groundwater flow and contaminant migration to the well. All groundwater flow between the upper and lower aquifers must pass through the aquitard by intergranular flow without preferential pathways, and this type of flow is extremely slow relative to flow in fractures or stratigraphic windows.

The discussion presented above focused on pathways through aquitards without consideration of the influence of the type of contaminant. The contaminant type is important because an aquitard can have excellent integrity for one type of contaminant and poor integrity for another. For example, the propensity for a dissolved contaminant such as nitrate (NO^-_3) to migrate through a fractured aquitard is generally much less than that for a chlorinated solvent, such as TCE, as a dense, immiscible liquid (i.e., DNAPL).

This report uses the term *aquitard integrity* in the general context of protecting aquifers, not just a particular well or wellfield. Aquitard integrity depends on three factors (Figure 1.3): (i) state of the hydrologic system (hydraulic head distribution), (ii) contaminant characteristics (dissolved, NAPL, particulate, microbial, reactive or degrading), and (iii) hydrogeologic characteristics (hydraulic conductivity, porosity, thickness, etc.). Furthermore, if human disturbances such as quarries or improperly constructed wells are present, they can change aquitard integrity by creating new pathways or short circuits for contaminant migration. Figure 1.4 illustrates interaction of the hydrologic system with improperly constructed wells and hydrogeologic characteristics of the aquitard. Discrete pathways, including fractures, faults, macropores, and human-induced pathways such as open boreholes, are found in some aquitards and can govern aquitard integrity. Chapter 3 of this report discusses other geologic pathways such as erosional or depositional "windows", facies changes, and subcrops.

Determining the degree of protection aquitards provide to underlying aquifers is a challenging task. Geologic information on its own is rarely, if ever, adequate and should always be complemented with other types of data. The literature describes many methods for investigating aquitards. While simple geologic criteria such as sediment or rock type is sometimes used in consideration of wellhead protection (U.S. Environmental Protection Agency 1991), information on hydrogeologic setting (Belitz and Bredehoeft 1990, Neuman and Neretnieks 1990, Simpkins et al. 1996), hydraulic head (Eaton 2002, Eaton and Bradbury 2003, Rophe et al. 1992) hydraulic conductivity (Shaw and Hendry 1998, Williams and Farvolden 1967, van der Kamp 2001) and hydrochemistry and isotopes (Hendry 1988; Hendry, Wassenaar, and Kotzer 2000; Nativ et al. 1995; Nativ and Nissim 1992; Pucci 1998; Pucci 1999; Ramenda, van der Kamp, and Cherry 1996; Stimson et al. 2001) is typically necessary for assessing aquitard integrity. Much emphasis concerning aquitards in the early hydrologic literature was directed at pumping tests with the goal of determining properties of both the pumped aquifer and the associated aquitard(s). However, determining the aquitard properties most relevant to contaminant migration based on interpretation of pumping tests in overlying or underlying aquifers is not possible if the drawdown data are obtained only from the aquifer (Lebbe and Van Meir 2000). Aquifer drawdown can reflect aquifer heterogeneity or uncertain lateral boundary conditions appearing to represent effects of flow through the aquitard. Analysis of changes in head data *within* aquitards while pumping overlying or underlying aquifers is more indicative of the vertical hydraulic conductivity of aquitards (Grisak and Cherry 1975, Neuman and Witherspoon 1972, Rowe and Nadarajah 1993).

Knowledge of the depositional and post-depositional environment of aquitards is essential for development of expectations of occurrence of windows or other preferential flowpaths. A laterally extensive aquitard can provide more or less protection of an underlying aquifer depending on the part of the aquitard under consideration. Near the edge of an aquitard, for example, an underlying aquifer may be semi-confined or unconfined because flowpaths can easily bypass the aquitard. As the aquitard thickens or deepens away from its periphery, an underlying aquifer may be increasingly protected. Surface geophysical (Crattie 1991; EddyDilek et al. 1997; Oatfield and Czarnecki 1991; Owen, Park, and Lee 1991; Young and Sun 1995) and geostatistical methods (Desbarats et al. 2001; James and Freeze 1993; Ritzi, Dominic, and Kausch 1996), have been applied to field data to assess the extent and continuity of aquitards, particularly in the case of shallow unlithified clay aquitards.

ASSUMPTIONS AND MISCONCEPTIONS

Various simplifications are common for the hydrogeology of aquitards. For example, textbook discussions involving aquifers and aquitards in the context of pumping tests are commonly restricted to consideration of idealized aquitards that are homogenous with uniform thickness and are laterally extensive (Figure 1.5). The hydrogeologic properties of aquitards are almost never homogeneous, and aquitards have non-uniform thickness. Because hydraulic measurements are rarely collected from within aquitards, analysis of water-level data from overlying and underlying aquifers is often assumed to give reliable insight into aquitard characteristics. Although this may be useful for assessment of aquitard leakage in an aquifer yield context, it is inappropriate in assessment of aquitard integrity. Contaminant migration through aquitards is often erroneously believed to depend only on bulk hydraulic properties of aquitards, without regard to preferential flowpaths in the aquitard or different contaminant types. Actual rates of contaminant transport through aquitards can be very different from those based on estimates of bulk flow rates. Using a two-dimensional, discrete-fracture model, Harrison, Sudicky, and Cherry (1992) showed even though the volumetric flow rates (i.e., Darcy flux) from an aquitard to an aquifer can be very low, contaminant transport through aquitards may be relatively rapid because of fractures, even very small fractures, if they fully penetrate the aquitard. Basic hydrogeologic techniques designed for aquifers, such as pumping and slug tests, commonly need modification to be appropriate for assessment of low permeability geologic media (Novakowski and Bickerton 1997, Shapiro and Greene 1995, van der Kamp 2001).

Two misleading terms common in the literature are *leaky* aquitards (aquitards which have significant water transmission), and *aquicludes* (geologic strata through which no water can pass). These terms come from a water supply perspective and are inappropriate for use in the context of contaminant migration. All aquitards are leaky in the sense that all aquitards transmit at least some (albeit often very small) quantities of water. Use of the term *leaky* implies two categories: some aquitards "leak" and others do not. Instead, there is a continuum of flow conditions, ranging from aquitards that transmit water at only small rates possibly imperceptible depending on the stress conditions or time scale of measurement, to those that transmit significant quantities of water.

While a conceptual model of a laterally extensive aquitard may be adequate for analysis of the hydraulic conditions created in the aquifer by pumping wells in a confined aquifer, aquitards are often not laterally continuous or uniformly thick over the entire area of relevance. Hydrogeologic properties of aquitards often vary over a much larger range than for aquifers because of fractures and preferential flowpaths. Aquitards can be more anisotropic than aquifers; hydraulic conductivity can be much higher in the horizontal than vertical direction due to stratification, which has important implications to interpretation of slug test data from monitoring wells. Conversely, vertical fractures may cause the vertical hydraulic conductivity component to greatly exceed the horizontal component. The proximity of the piezometers in the aquitard to the fractures will influence the detectability of these vertical pathways.

Even though flow rates per unit area (specific discharge or water flux) through aquitards can be low, the total volumes of flow summed over the entire extent (possibly hundreds of square kilometers) of an aquitard can amount to a major fraction of water pumped from an underlying aquifer (Bredehoeft, Neuzil, and Milly 1983; Neuzil 2002; Walton 1960). This can be due in part to the large aquitard storage coefficients (storage coefficient is the volume of water released per unit area of formation per unit change in head). While conventional hydrogeologic analyses assume predominantly vertical flow in aquitards, measurements from wells finished in bedrock aquitards have shown considerable lateral flow (Eaton and Bradbury 2003, Peffer 1991). Therefore, determining the internal characteristics of aquitards and their natural complexities is commonly essential for better understanding of the groundwater flow paths extending from contaminant source areas to wells situated beneath aquitards.

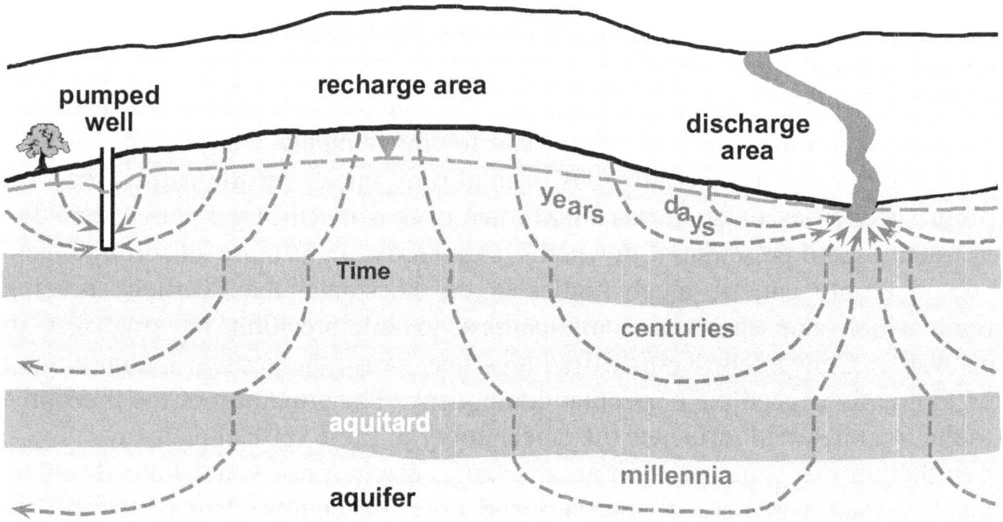

Source: Adapted from Tóth 1995. Hydraulic continuity in large sedimentary basins. *Hydrogeology Journal* 3:4-16. With kind permission of Springer Science and Business Media.

Figure 1.1 Schematic diagram showing groundwater flow lines in a regional system comprised of laterally extensive aquifers and aquitards. Slow groundwater flow through the aquitards (unfractured) results in much older water in successively deeper aquifers

(a)

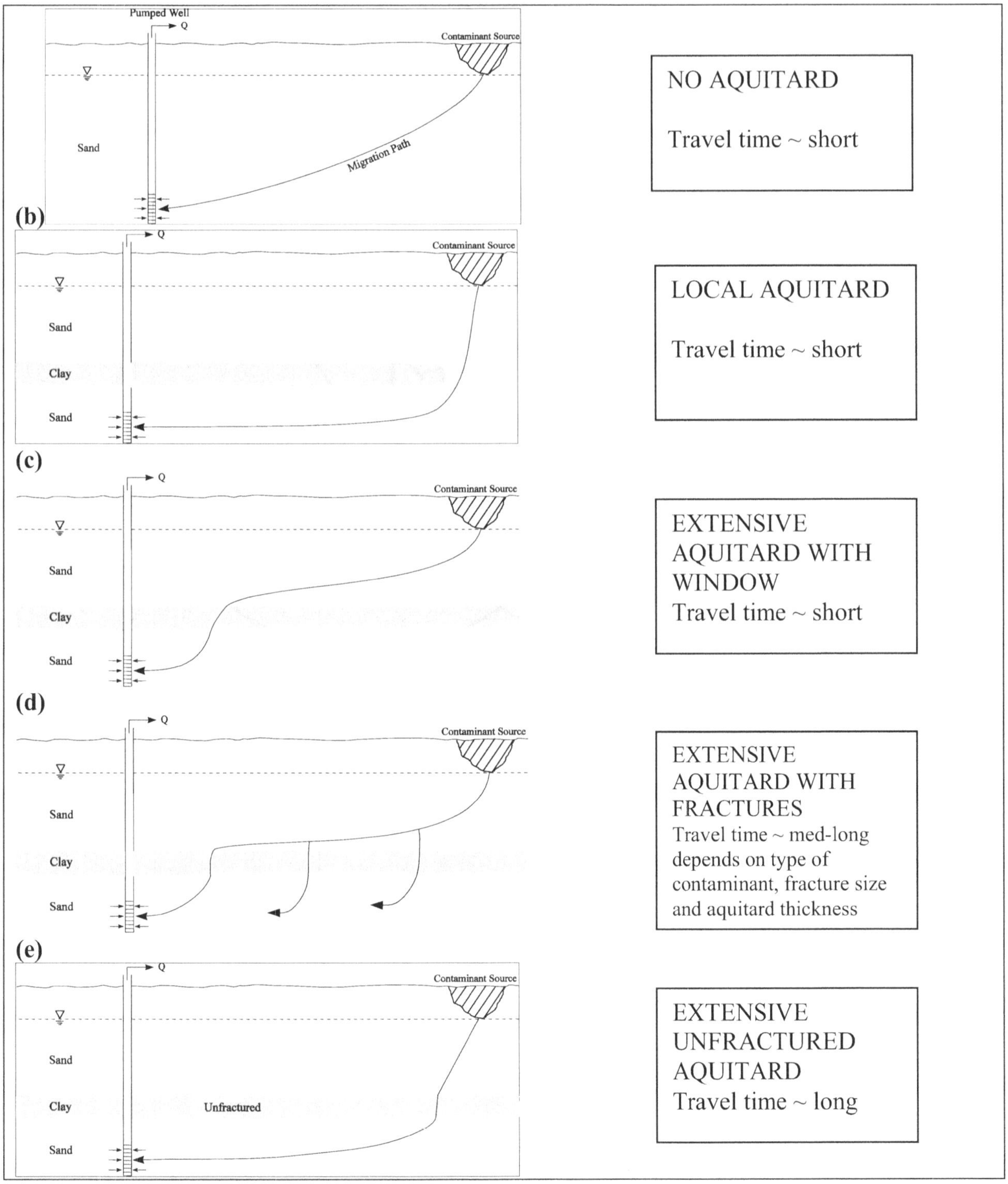

Figure 1.2 Conceptual pathways for contaminant migration and travel times along the pathways from a near surface source to a pumping well: influence of aquitards with different features

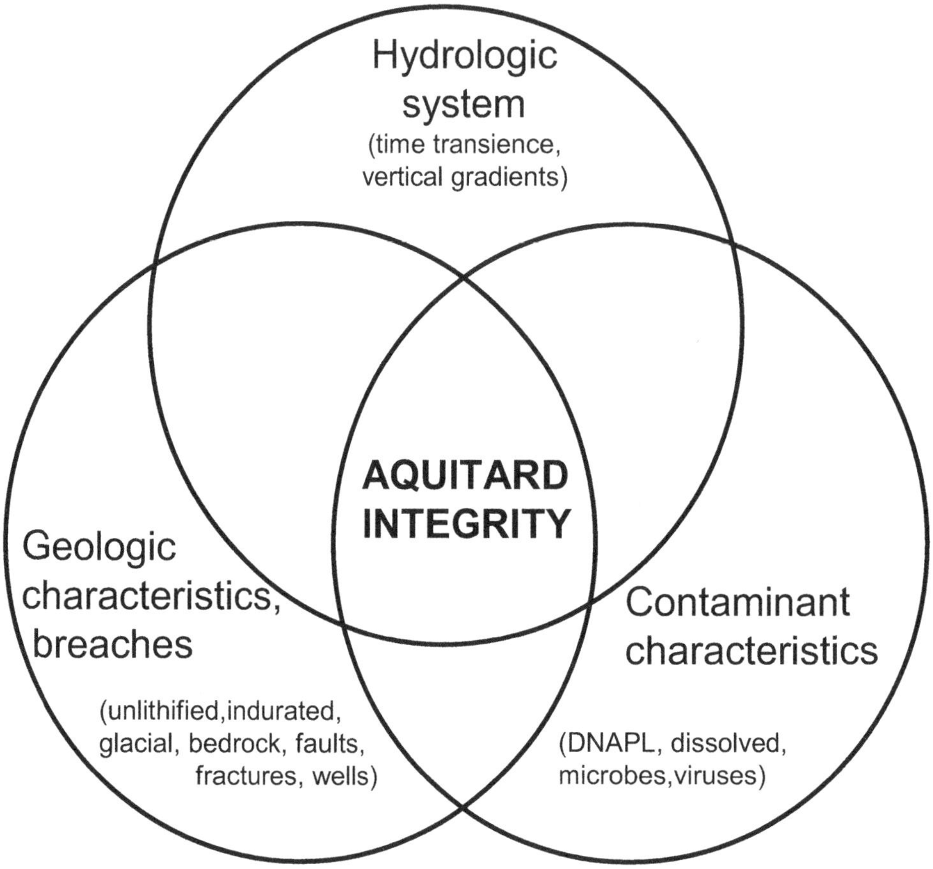

Figure 1.3 The three components of aquitard integrity

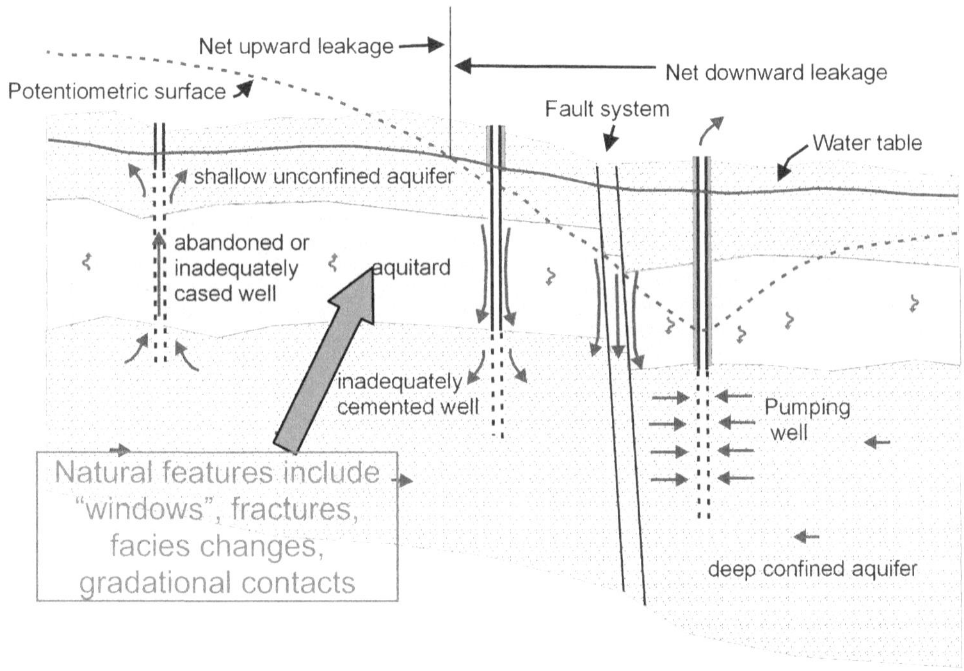

Source: Adapted from U.S. Environmental Protection Agency 1987, U.S. Environmental Protection Agency 1991

Figure 1.4 Examples of potential natural and human-induced groundwater flow paths through a hypothetical aquitard. In areas where the potentiometric surface of the deep confined aquifer is above the water table there is a potential for upward flow across the aquitard. In areas where the water table is higher than the potentiometric surface there is a potential for downward flow across the aquitard. Discrete features such as fractures, faults, and existing wells can provide conduits for rapid movement of water and/ or contaminants.

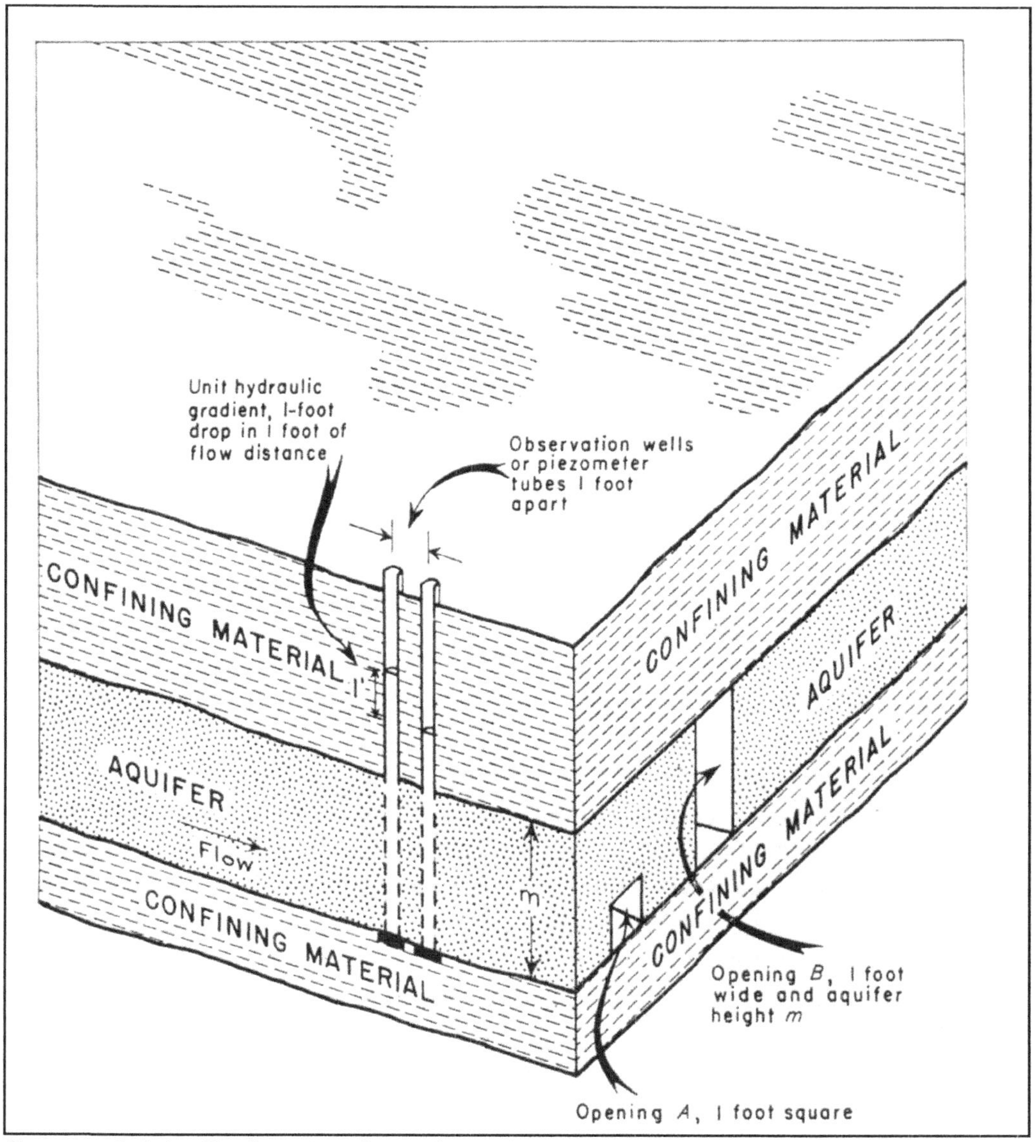

Source: Originally from Ferris et al. 1962

Figure 1.5 Common textbook illustration of aquifers and confining beds (aquitards), used for consideration of pumping tests

CHAPTER 2

HYDROGEOLOGIC SETTINGS FOR AQUITARDS

The basic geological characteristics of different types of aquitards, including their morphology, geologic origin, and history, impart their hydrogeologic properties. Understanding the range of possibilities for hydraulic properties and occurrence of contaminant pathways in any individual aquitard is important for assessing aquitard integrity and the vulnerability of the underlying aquifer. Identification of the geologic origin and post-depositional history can focus analysis on the most likely relevant features (e.g., eliminate study of geologically unlikely pathways), and allow knowledge obtained from study of similar aquitards elsewhere to be applied. However, site-specific assessment of the hydraulic characteristics is essential for development of reliable understanding of the aquitard's role in protecting the underlying aquifer. Such assessments almost always require depth-discrete monitoring and sampling to provide the critical information for pathway identification.

BASIC CHARACTERISTICS OF AQUITARDS

Aquitards can be classified on several different levels. One simple distinction is between surficial and buried aquitards (Figure 2.1). A second basic distinction is whether the aquitard is unlithified (i.e., non-indurated), formed of clay and silt, usually deposited by glacial, lacustrine or marine environment processes or is indurated material such as shale (Figure 2.2). Aquitards typically found at or near land surface are nearly always unlithified while aquitards important to bedrock aquifers are part of the bedrock or part of the overburden. Knowledge gained from the study of unlithified aquitards can be applied to bedrock aquitards and vice-versa, but in some cases such knowledge transfers are not appropriate.

Fractures, macropores or other pathways through aquitards can diminish their capability to protect underlying aquifers from contamination. For example, open fractures can greatly reduce aquitard integrity if the fractures fully penetrate the aquitard (Figure 2.3a). In general, thicker aquitards provide more protection than thinner ones (Figure 2.3b), but in some cases open fractures occur throughout thick aquitards.

Geologic Origin

Aquitards exhibit a wide array of geologic lithologies, including silts, clays, shales, mudstones, siltstones, carbonates, well-cemented sandstones, quartzites, evaporates, and igneous and metamorphic rocks. The lithology or mineral composition of aquitards covers a range of materials, including siliceous volcanic lava flows (Larson et al. 2000, Li 1991, Updegraff et al. 1981), shale or siltstone (Eaton and Bradbury 2003; Lindgren 2001; Neuzil 1988; Williams, Teodoru, and Mills 1996), chalk (Nativ and Nissim 1992, Rophe et al. 1992), dolomite (Conlon 1998), cemented sandstone (Hornung and Aigner 1999), clay and silt-rich Pleistocene till (Grisak and Cherry 1975; Keller, van der Kamp, and Cherry 1986; Shaw and Hendry 1998), lacustrine deposits (Brownell 1986; Harloff 1942; Husain, Cherry, and Frape 2004; Oretega-Guerrero, Cherry, and Aravena 1997), marsh sediments (Döll and Schneider 1995), marine clays (Castle and Miller 2000, Desaulniers and Cherry 1988, Pucci 1998) and even lignite (Geistlinger et al. 1998). Lithology or mineral composition alone is not sufficient to identify unique characteristics that may determine aquitard integrity. For example, aquitards with quite different hydrogeological properties, such as unlithified Pleistocene till and indurated shale bedrock, may have similar lithologic compositions. Furthermore, aquitard properties may depend on specific beds that form only part of a geologically-defined stratigraphic formation. Mechanical properties (Rijken and Cooke 2001) that determine fracturing when the aquitard is subject to tectonic or weathering forces may also be important. In some regions weathering results in sufficient disaggregation of bedrock aquitards (McKay, Sanford, and Strong 2000; Nativ, Halleran, and Hunley 1997; Schreiber, Moline, and Bahr 1999) to cause them to have characteristics similar to unlithified aquitards. We therefore use geologic origin, encompassing both lithology and postdepositional processes, to classify aquitards and as a guide to hydrogeologic properties.

Post-depositional alteration of aquitards can form pathways for groundwater flow or contaminant transport. For example, aquitards in overlying carbonate bedrock are vulnerable to disruption caued by bedrock dissolution resulting in sinkhole formation. Land surface collapses of this type have been documented in Illinois, undermining a valley-fill aquitard (Panno et al. 1994), and in the southeastern United States (Jancin and Clark 1993). In the latter case, the Hawthorn Group, a major aquitard of mixed lithology (silt, clay, sand) confining the regional Floridan carbonate rock aquifer, was breached. Preferential vertical flow through lakebeds in basins formed in this manner has been documented with water balance studies (Motz, Sousa, and Annable 2001; Watson, Motz, and Annable 2001).

Heterogeneity and Morphology

Many aquitards exhibit thickness variations or major internal lithologic variations (Figure 2.4). Aquitards composed of individual lenses of clay materials, or mudrocks containing sandstone interbeds or transmissive bedding fractures, can be conveniently viewed as single low-permeability units for many hydrologic applications. They impede flow between aquifers and cannot be pumped for water supply except in rare circumstances. In other cases, transmissive interbeds or bedding fracture zones exist within aquitards forming mini-aquifers with aquitard above and below. Wells with open intervals across aquitard-aquifer sequences may produce major quantities of water (Bishop et al. 1993a, Bishop et al. 1993b).

Knowledge of aquitard extent and thickness is critical to assessing aquitard integrity. The geologic history of a hydrostratigraphic unit controls these features. Figure 2.5 shows a series of conceptual aquitard settings, and illustrates how depositional history, erosion, faulting, fracturing, and other processes control morphology. Crystalline bedrock sometimes forms a local low-yielding aquifer due to fractures. However, zones with minimal fracture permeability in crystalline rock can be aquitards; this type of aquitard occurs in the U.S. Appalachian plateau (Peffer 1991), Newark basin (Maguire 1998, Michalski and Britton 1997), Nebraska (Barrash 1986, Barrash and Ralston 1991), and Great Britain (Bishop et al. 1993a, Bishop et al. 1993b). Settings where low hydraulic conductivity rocks are important primarily as hydrocarbon sources or traps at great depth, but at shallow depth are important in the groundwater flow system, occur in regions such as the Green River Formation on the U.S. Colorado Plateau. A lithological subdivision of this formation called the Mahogany zone, with low vertical hydraulic conductivity causes artesian conditions in fractured shale (Taylor and Hood 1988).

GEOLOGIC ORIGINS OF AQUITARDS

Geological formations that form aquitards fall generally into three classes of geologic origin: (i) common unlithified aquitards (clay, till, lacustrine sediment, fine-grained alluvium, etc.); (ii) common indurated sedimentary aquitards (shale, mudstone, siltstone, limestone, dolostone, etc.); and (iii) less common aquitards (evaporites, volcanic rocks, intrusions, metamorphic rocks, etc.). In each category, geologic depositional and post-depositional (diagenetic) processes greatly influence the development of possible preferential flowpaths which affect qualities that provide some degree of aquifer protection.

Unlithified Fine-Grained Sedimentary Aquitards

Aquitards in this category include till, glaciolacustrine sediments, non-glacially deposited lacustrine sediments, and fine-grained alluvial, estuarine or overbank deposits. Depending on their clay mineral and water contents, they may have considerable plasticity and may not have open fractures at depth much below the water table. However, where these deposits occur at the surface, weathering promoted by drying and wetting cycles and/or freeze-thaw cycles combined with geochemical process cause fractures typically open in the vadose zone and also open below the water table to maximum depths depending on local circumstances. Figure 2.6 shows fractures of this origin in the extensive thick glaciolacustrine aquitard in southwestern Ontario described by Husain, Cherry, and Frape (2004). Aquitards with any of these depositional origins may have discontinuities or windows where they have been eroded or disturbed by human activities (Desbarats et al. 2001).

Most important aquitards in North America are formed of mud or mudstone and the origin of these deposits are described in exceptional clarity for a wide readership in the recent textbook by Potter, Maynard, and Depetris (2005). This book shows numerous x-radiographs of cores taken from several types of mud deposits. X-radiography is a method that displays the internal structure of cores with clarity of structural detail much beyond that of any other method for core examination. The x-radiography method for examination of core samples of Unlithified sediment was pioneered by Coleman and Gagliano (1965) and has been used in most prominently in studies of mud deposits in the Gulf Coast Region of the United States. X-radiography is relevant in investigations of aquitard integrity because it can, in some cases, assist with identification of the geologic origin of an aquitard and in the determination of the presence or absence of secondary features such as fractures, root channels or structures caused by organism activity and these secondary features may have implications concerning pathways for groundwater flow and contaminant migration.

Glacial Sediment

Diamicton deposited by Pleistocene glaciers (i.e., till) consists of material deposited underneath and in front of advancing glaciers or from melting ice sheets. These modes of deposition generally result in poorly sorted material containing sand, cobbles and boulders in a finer-grained matrix. Where these deposits contain abundant clay or silt, they commonly form aquitards overlying or underlying sand or gravel aquifers (Rodvang and Simpkins 2001, Simpkins and Bradbury 1992). The complexity of glacial deposition can result in preferential flowpaths within these aquitards (Gerber, Boyce, and Howard 2001; Gerber and Howard 1996;

Sminchak, Dominic, and Ritzi 1996). Hydrogeologic properties have strong dependance on till genesis (Haldorsen and Kruger 1990). Boundaries of such deposits are commonly associated with moraines (ridges in the landscape) marking the farthest advance of ice sheets.

Mechanical stresses during and after deposition can have important effects on the integrity of till aquitards. If the till is deposited under a continental ice sheet, subsequent melting or withdrawal releases the till from the mass of the overlying ice. Resulting expansion of the till sediment or rebound of the earth's crust can create a network of fractures, increasing the effective bulk hydraulic conductivity of the sediment (Figure 2.6) (Bradbury et al. 1985, Simpkins and Bradbury 1992). Fracturing intensities can vary between physiographic provinces (Brockman and Szabo 2000).

Biopores and desiccation fractures can increase effective hydraulic conductivity and create preferential flowpaths in the upper few meters of the till (Jorgensen and Fredericia 1992, McKay and Fredericia 1995). Later readvances of the continental ice sheets and additional mechanical stresses can cause secondary fracture sets (Klint and Gravesen 1999; Sidle et al. 1998). Due to these different mechanisms, there may be multiple orientations of fracture sets (Brockman and Szabo 2000; Haldorsen and Kruger 1990; Helmke, Simpkins, and Horton 1996; Helmke, Simpkins, and Horton 1997), which may or may not be interconnected. Where fractures are open and form interconnected pathways from the top to bottom in the aquitard, contaminants transported through the fracture networks can enter aquifers (Broholm et al. 1999a; Broholm et al. 1999b; Harrison, Sudicky, and Cherry 1992; Schmidt 1993). Weathering processes after deposition can enhance existing fractures near the land surface, seal previously open fractures by precipitation of dissolved minerals, or merely coat fracture surfaces (Corrigan, Jamieson, and Remenda 2001).

Some regional aquitards are formed of surficial glaciolacustrine sediment deposited in lakes formed in front of continental ice sheets (Desaulniers and Cherry 1988; Husain, Cherry, and Frape 2004; Remenda, Cherry, and Edwards 1994). Glaciolacustrine aquitards are typically better sorted and more stratified than till, and in some cases have a significant sand component (Clayton and Attig 1989) resulting in a higher local bulk hydraulic conductivity. In many cases the sand occurs in horizontal layers imparting higher conductivity in the horizontal direction. The extent of surficial glaciolacustrine sediment can often be inferred by geomorphic analysis of landscapes. Different phases of deposition of glaciolacustrine sediments can result in interbedded sandy aquifer materials and silt or clay beds that form aquitards (Brownell 1986). These aquitard beds may be relatively thin (0.5 m) and poorly recognized but hydrogeologically important (Harloff 1942) if they contain at least 10-20% clay or silt content. In some cases, glaciolacustrine clay forms numerous disconnected beds and lenses of various sizes rather than a continuous clayey layer (Harloff 1942), with the finest-grained zones nearest the center of depositional basins. Glaciolacustrine sediments are subject to the same weathering processes as glacial till aquitards where they are exposed at land surface. Surficial clayey or silty aquitards that have not been overridden by re-advancing glaciers have abundant fractures at shallow-depth; at greater depths, fully penetrating open fractures in thick aquitards can be absent, particularly if the porosity and plasticity are appreciable. Hydraulic properties can depend on whether aquitards have been exposed to the land surface in their geologic past, even if they have since been buried by glacial or other sediment.

In addition to fractures, other features of Pleistocene deposits can form preferential flowpaths in aquitards. For example, eskers are narrow, sinuous ridges of coarse-grained material formed under glacial ice, and can be embedded in clay or silt-rich till sheets. Ice-wedge casts (Dansart 2001) are predominantly vertical deposits of sand or wind-blown silt formed during freeze-thaw cycles in former permafrost terranes. Subhorizontal shear joints (Haldorsen and Kruger 1990) or microbeds of sand (Morrison, Parker, and Cherry 1998), not detectable using conventional well-drilling methods, as well as sand lenses (Gerber, Boyce, and Howard 2001) can be important lateral pathways for contaminant transport in clay- or silt-rich glacial materials. In well sorted and layered glacial sequences, the combination of such lateral high–conductivity beds and occasional vertical interconnections can create a stair-step pattern of preferential flowpaths for contaminant transport (Sminchak, Dominic, and Ritzi 1996) and DNAPL flow (Morrison, Parker, and Cherry 1998). Due to variations in small-scale depositional processes or erosional removal of aquitard material after deposition (Desbarats et al. 2001; Dominic, Ritzi, and Kausch 1996), aquitards formed by glacial sediments should not be assumed devoid of windows. Identification of the depositional framework of glacial deposits can provide insight into the internal complexity relevant to hydraulic conductivity distribution (Anderson 1989).

Non-glacial Sediment

Non-glacial, unlithified aquitards can form in environments where fine-grained sediment is deposited, such as estuaries, wetlands, fluvially-deposited systems including deltas, closed lacustrine basins, and marine settings or continental shelves. Wetland sediments constituting aquitards have been studied in Florida (D'Amore et al. 2000) and the Netherlands (Döll and Schneider 1995). Heterogeneity in these marshy sediments creates challenges for evaluating effective hydraulic properties at different scales (Bierkens 1996). In fluvial systems, broad lateral and vertical variability in sediments depends on the evolution and type of river deposition. Clay-rich deposits collect in abandoned meander channels, called oxbows, and can form discontinuous aquitards of limited lateral extent. In some regions, alluvial aquifers are overlain by fine-grained overbank sediments (Aslan and Autin 1996, Hanor 1993) acting as aquitards. After deposition, some of these aquitards are breached by downcutting channels (EddyDilek et al. 1997). Aquitards in alluvial fans (Stimson et al. 2001, Timms and Acworth 2002) and deltas (Manzano, Custodio, and Jones 1990) can be complex and made up, for example, of laterally extensive but thin paleosols (Weissmann and Fogg 1999).

The hydrogeology of unlithified lacustrine deposits in the Mexico City area has been studied by (Ortega-Guerrero and Farvolden 1989; Ortega-Guerrero, Cherry, and Rudolph 1993; Ortega-Guerrero, Rudolph, and Cherry 1999; Rudolph 1989) and others. The high porosity and compressibility of these deposits has resulted in major land subsidence problems due to pumping from the underlying aquifer, (Rudolph and Frind 1991). Paleo-evaporative processes can enrich the salinity of porewater in such lacustrine aquitard sediments (Ortega-Guerrero, Cherry, and Aravena 1997), which presents a potential negative impact of aquifer water quality as the saline porewater is expulsed from the aquitard due to consolidation (Rudolph, Cherry, and Farvolden 1991). The distribution of natural water chemistry in clayey aquitards can indicate processes governing solute transport (Ortega-Guerrero 2003, Pucci 1998, Pucci 1999). Pucci, Ehlke, and Owens (1992) describe how aquitards affect aquifer water quality in the New Jersey Coastal Plain.

Aquitards can also form on continental oceanic shelves (Castle and Miller 2000) and coastal plains (Erskine and Fisher 2002), and may have complex geometry depending on their depositional histories. Aquitards can control the position and depth of the saltwater-freshwater interface (Back 1985). In coastal areas, salinity in aquitards and adjacent aquifers can also result from seawater intrusion (Mas-Pla et al. 1999).

Where lithological heterogeneity such as sand lenses or erosional windows are not a cause of preferential pathways in clay-rich, unlithified aquitards, fractures are the most probable pathway for preferential flow. A generalized, conceptual model for fracture distribution and orientation formed due to contraction caused by dessication and weathering processes emerged from studies in pits excavated in surficial clayey aquitards (Brockman and Szabo 2000, McKay and Fredericia 1995) (Figure 2.7). Although this conceptual model was developed for surficial till and glaciolacustrine aquitards, it is also applicable to non-glacial lacustrine or other unlithified aquitards exposed at ground surface. The dominant fractures are vertical, but horizontal fractures can also be present. The density of fractures and preferential pathways typically decreases with depth below the oxidized zone, but can extend to 10-15 m below the water table and perhaps deeper. If the aquitard is composed of different sedimentological units, fractures in deeper units may have formed when these units were exposed at surface and then remained open after burial.

Aquitards in Sedimentary Bedrock

Aquitards formed in indurated bedrock behave in a similar fashion to unlithified aquitards, but relatively few studies have characterized the hydrologic properties of these rocks. Usually they are defined in contrast to adjacent aquifer systems (Back, Rosenhein, and Seaber 1988), and minimal hydrogeologic data are available. Geologically, bedrock aquitards are typically heterogeneous, composed of numerous interbedded lithologies, including shale, siltstone, mudstone, highly cemented sandstone, dolomite, limestone, marl and chalk. Most published information on lithified aquitards pertains to sedimentology, paleontology or petroleum geology because these geologic units can form source beds or "cap rock" (seals) for hydrocarbon resources.

Detailed lithologic characterization is important in the evaluation of hydrogeologic properties of indurated aquitards. For example, the Eau Claire Formation in Wisconsin (outcrop shown in Figure 2.8) is predominantly sandstone, but performs as an aquitard only where it contains a shaley zone (Bradbury et al. 1999). In eastern Wisconsin, and in adjacent states, the Maquoketa Formation, an Ordovician shale, forms a regionally important aquitard (Figure 2.9) and is usually reported to be at least 50 m thick. However, the hydraulic properties in the formation are not uniform with depth. Recent field studies show the shale-rich basal zone provides nearly all of the vertical resistence to flow (Eaton 2002) but this zone is only a small part of the whole formation (Figure 2.10). The Tunnel City Formation is considered a regional aquitard in Wisconsin, but numerous springs issue from its bedding-plane fractures (Swanson 2002). This is an example illustrating the importance of distinguishing horizontal from vertical hydraulic conductivity as well as keeping in mind the scale and context at which the hydrogeologic unit is assessed.

Controls on Fracture Development in Bedrock Aquitards

Weathering features or secondary fracture porosity can be important with respect to the properties of a rock unit (Runkel et al. 2003). Mechanical properties of beds of different lithology influence development of fractures in sedimentary bedrock and control the vertical extent of fractures (Dutton et al. 1994, Rijken and Cooke 2001). The greater the lithological heterogeneity, the more likely vertical fractures are offset (Cooke and Underwood 2001, Pollard and Aydin 1988). Vertical fracture or joint density differs according to bed thickness and resistance (Gross et al. 1995, Helgeson and Aydin 1991, Narr and Suppe 1991). Small, vertically-oriented fractures or joints in more indurated beds often are offset by ramps (Dutton et al. 1994), or much gentler dipping and curved segments (Russell and Harman 1985) in softer mudrock, shale or marl beds. Hence, interconnected, throughgoing preferential flowpaths for contaminant transport in thick bedrock aquitards may be absent even though abundant fractures occur.

Data from bedrock (Eaton 2002, Maguire 1998, Michalski and Britton 1997, Pfeffer 1991) indicate rock aquitards are highly anisotropic, with the most extensive fractures commonly orientated along bedding planes with shorter but, in the context of aquitard integrity, often equally or more important fractures (e.g., joints) in the near-vertical orientation (Figure 2.2). This depiction of the fracture network is much different from the generalized conceptual model for fractures in unlithified clay deposits (Figure 2.3). The common assumption of primarily vertical flow in aquitards may not be appropriate for some lithified aquitards where complex flow paths influenced by bedding planes linking vertical fractures may be dominant (Eaton and Bradbury 2003).

Post-depositional Processes

Bedrock aquitards often have undergone tectonic stress causing hydraulic property changes, structural complexity or deformation. Deep burial over geologic time scales reduces the primary porosity (i.e., matrix porosity) of bedrock aquitards. Subsequent uplift, folding, or tectonic or glacial unloading can cause fracture formation within and across beds or along bedding-plane partings. The integrity of aquitards can be breached by normal faults with displacement greater than aquitard thicknesses, juxtaposing them with adjacent aquifers (Barker, Bush, and Baker 1994). Fault zones in brittle rocks can have very complex and heterogeneous structure (Caine, Evans, and Forster 1996; Smith, Forster, and Evans 1990). The relative permeability of fault zone cores (Antonellini and Aydin 1994) may be quite different from that of adjacent damage zones. The dynamic history of the fault may lead to enhanced fracture frequency near faults, potentially providing interconnected pathways for contaminant transport, or enhanced local hydraulic conductivity (Sen and Abbott 1991). Fracture observations in aquitard rock core and outcrop (Dutton et al. 1994) indicate although fracture swarms can occur anywhere, their intensity and interconnectedness is greater near master fault zones and structural folds. However, associated fluids can also seal these fractures by cementation over geologic time (Nelson et al. 1999). Conversely, in less deformed bedrock aquitards in carbonate rock (limestone, dolomite or chalk) dissolution processes enhance the development of preferential flowpaths along bedding-planes and other fractures (Nativ and Nissim 1992) compared to clastic (detrital shale, mudstones or siltstones) lithologies. In poorly-lithified sediments, fault zones are not extensively fractured and generally do not form vertical flow conduits, but restrict horizontal flow (Rawling, Goodwin, and Wilson 2001).

Other Types of Aquitards

Evaporates and volcanic rocks can also form aquitards. Evaporites consist largely of halite ($NaCl$) and associated anhydrite ($CaSO_4$), which form soluble rocks with low permeability.

In many sedimentary basins, including the Michigan Basin and U.S. Gulf Coast, they form stratigraphic seals for hydrocarbon resources. For example, an evaporite aquitard (Kreitler et al. 1985) underlies the well-known High Plains (Ogallalla) aquifer. Hydrogeologic properties of evaporites are known from studies at waste repository sites, such as the WIPP site in New Mexico (Beauheim and Roberts 2002). Although such aquitards are important in regional hydrogeology, they are not relevant to protection of drinking water supplies, because underlying aquifers tend to be saline.

Aquitards occur in volcanic rocks, but are difficult to identify. Volcanic rock sequences form major aquifers in the Columbia Lava Plateau (Lindholm and Vaccaro 1988), and Hawaii (Hunt, Ewart, and Voss 1988), and also on the Indian subcontinent. The hydrology of basalt aquifers and aquitards has been studied in Idaho (Li 1991, Updegraff et al. 1981) and the state of Washington, U.S. (Larson et al. 2000). The internal structure of lava flows in which different horizons are more or less fractured or rubbly determines the primary paths of groundwater flow. Major transmissive fracture zones in lava interflow zones are considered aquifers and can be confined by aquitards formed ofless fractured basalt (Spane and Thorne 1985). However, vertical cooling fractures in these aquitard units can provide preferential pathways for flow and contaminant migration. Basalt aquifer sequences in Hawaii are more uniform than those found in the Columbia and Snake River basalts in the northwestern U.S. and are commonly interbedded with clays, silts or pyroclastic materials that also form aquitards (Leonhart and Hargis 1990).

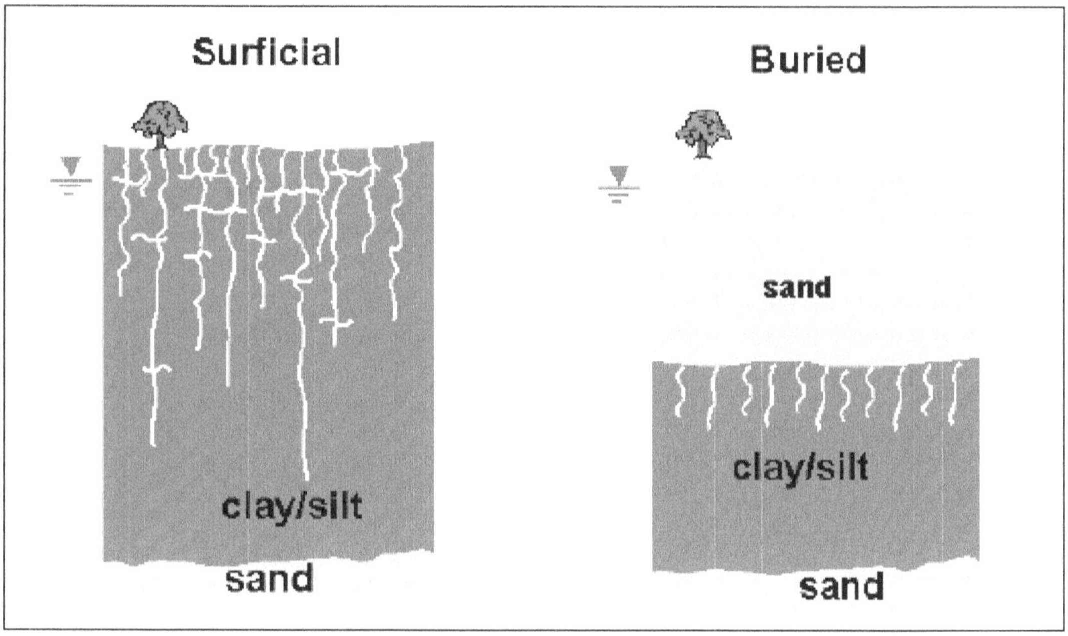

Figure 2.1 Two general types of non-indurated clayey or silty aquitards: surficial and buried.

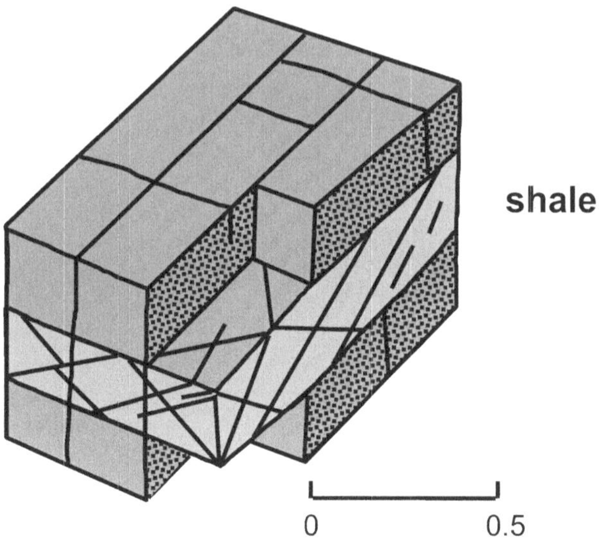

Source: Chernyshev and Dearman 1991. Reprinted from *Rock Fractures,* Chernyshev, S.N. and W.R. Dearman. Page 11. Copyright 1991 with permission from Elsevier.

Figure 2.2 Illustration of shale bed within sandstone. The frequencies and apertures of fractures are typically different from the sandstone to shale

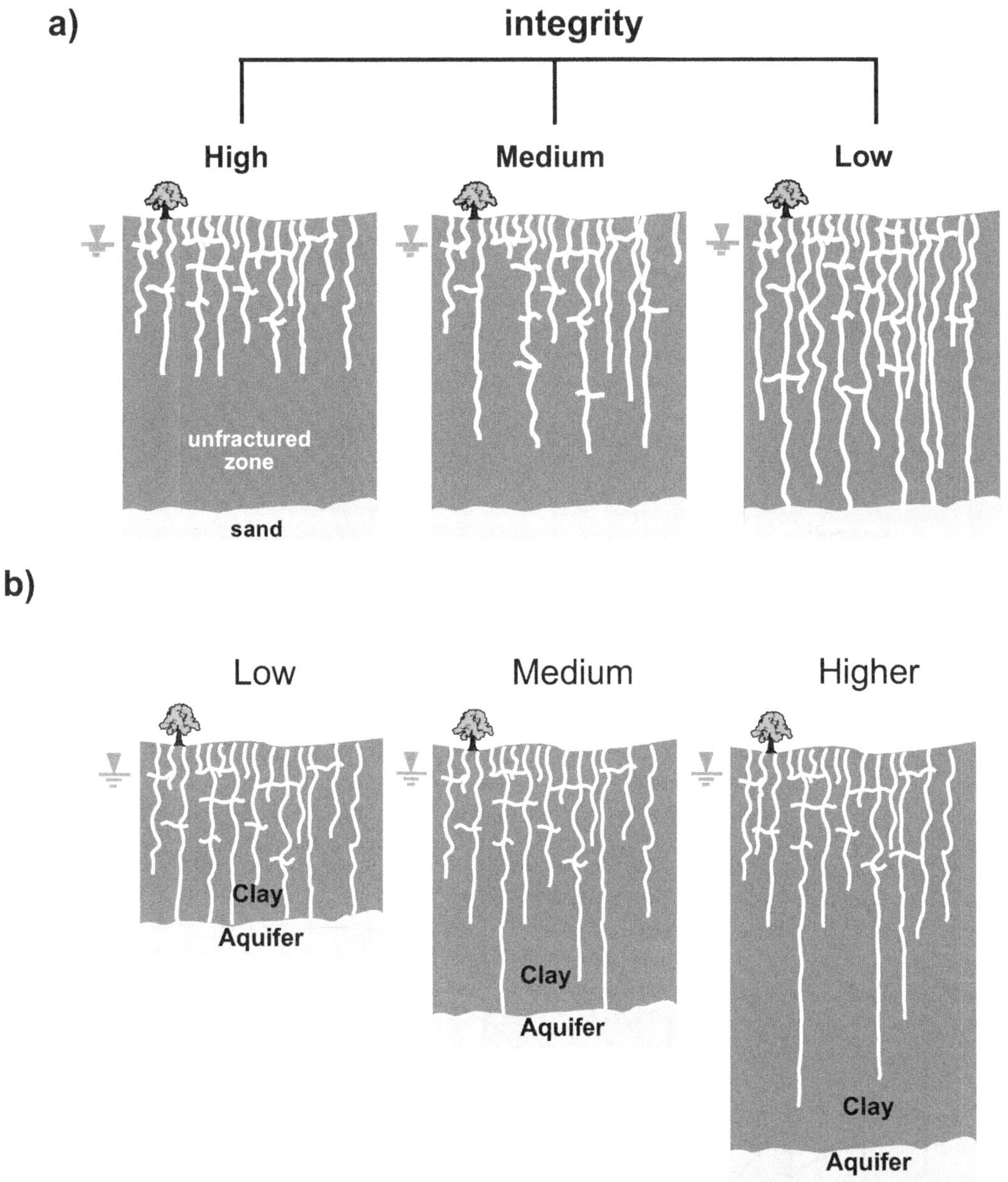

Source: Parker and Cherry, University of Waterloo, Earth 653 Course Notes

Figure 2.3 Aquitard integrity is commonly influenced strongly by: a) maximum depth of open fractures, and b) aquitard thickness

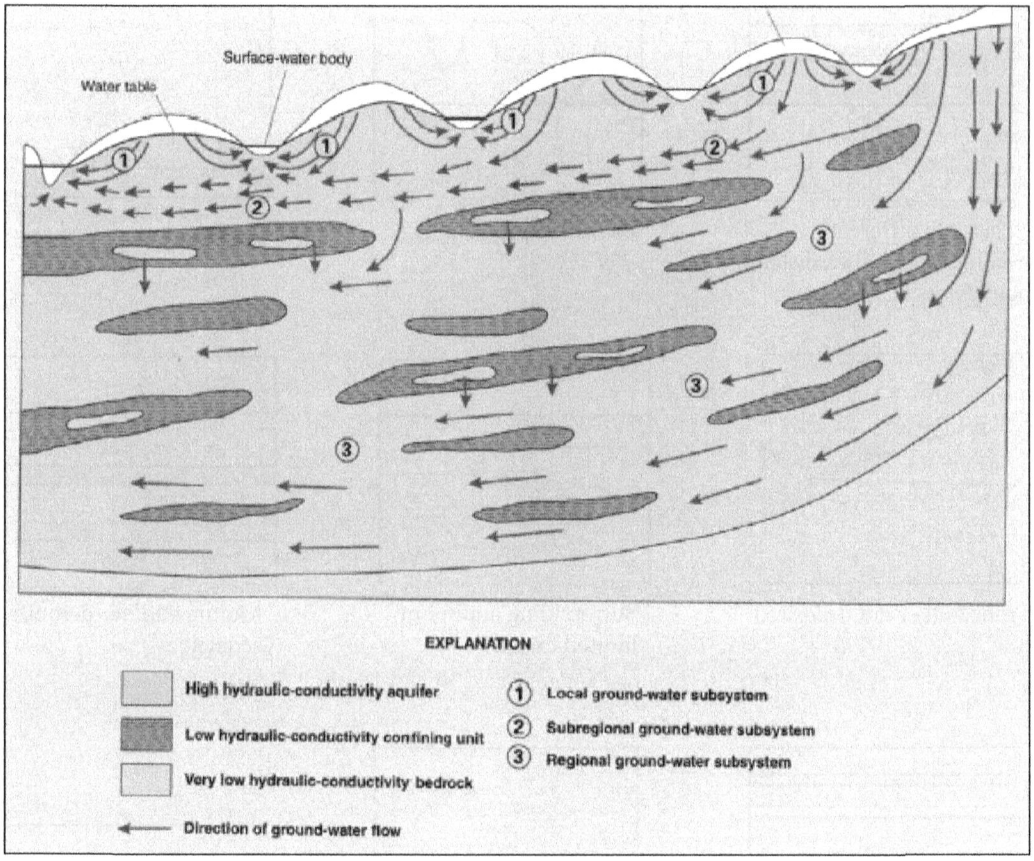

Source: Alley, Reily, and Frank 1999

Figure 2.4 Aquitards are often discontinuous on a regional scale

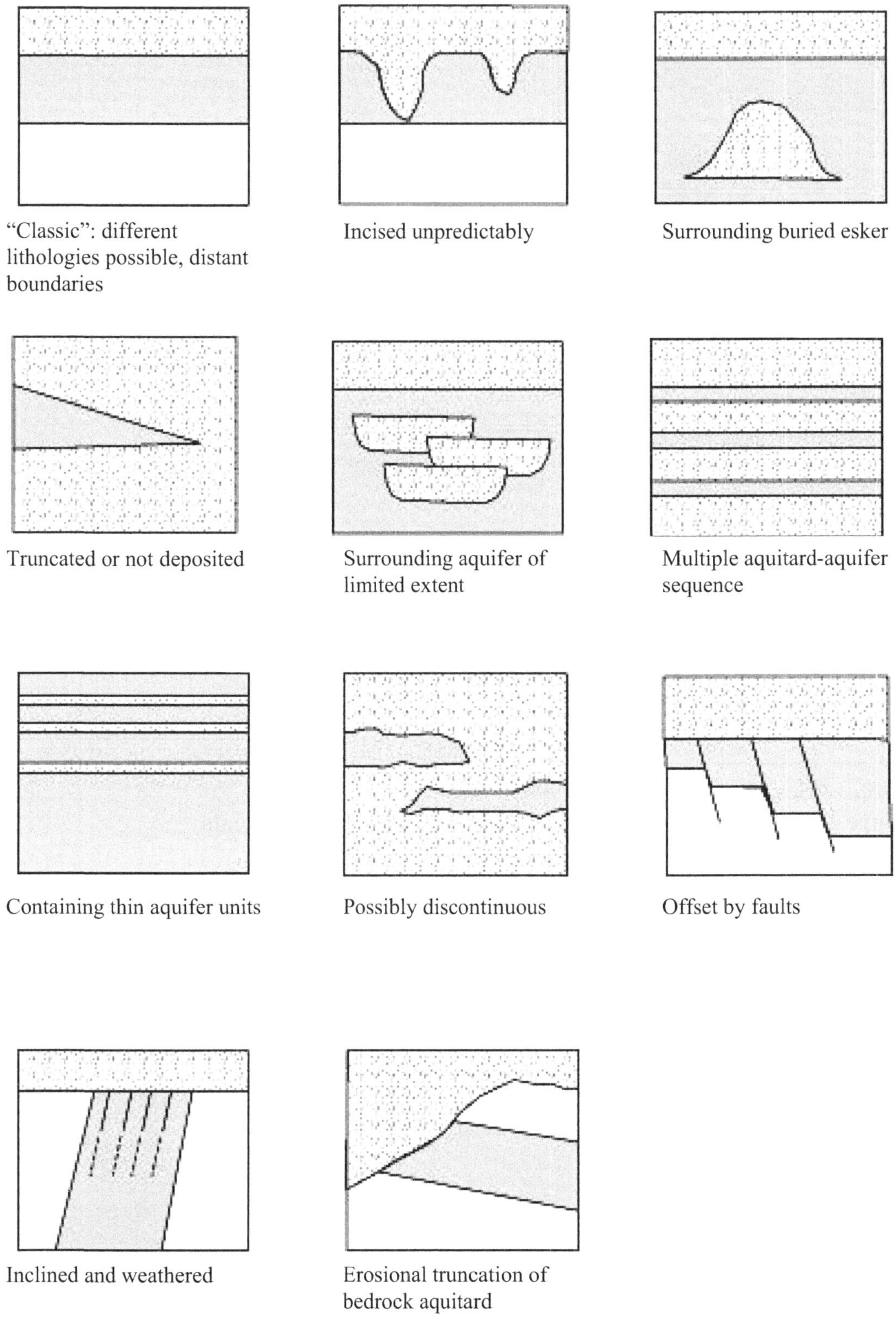

Figure 2.5 Several conceptual models of aquitards due to variations in depositional settings and postdepositional processes. Grey shading represents aquitard

Figure 2.6 Photograph of fractures in a surficial clayey aquitard

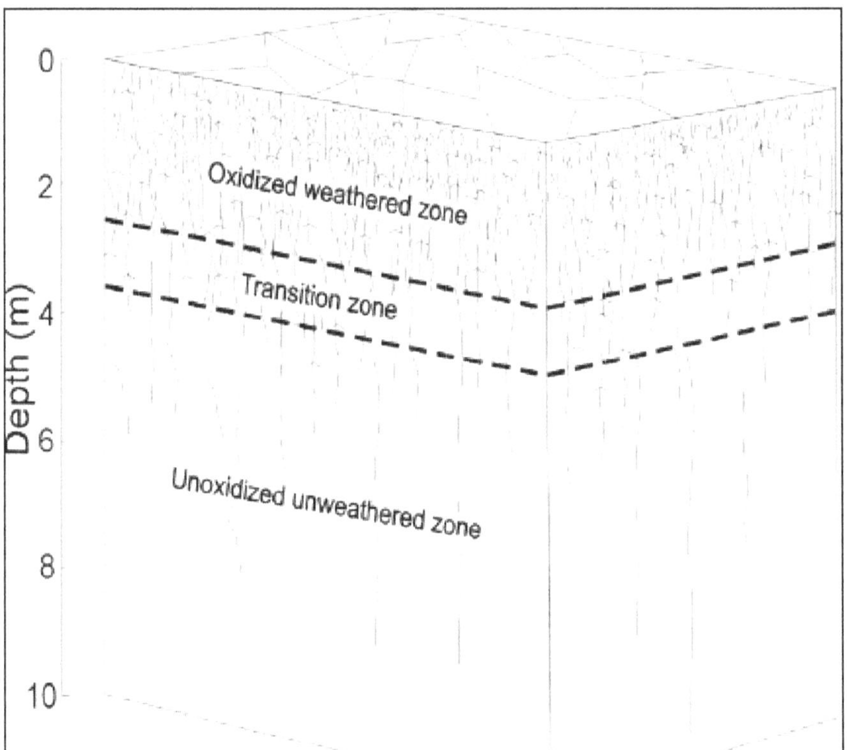

Source: Adapted from McKay and Fredericia 1995

Figure 2.7 Conceptual model of fracture distrubution in surficial clayey aquitard

Figure 2.8 Outcrop of the shaley part of the Eau Claire Formation, an important regional bedrock aquitard in south-central Wisconsin. Inset shows heterogeneity at a small scale (note pen for scale)

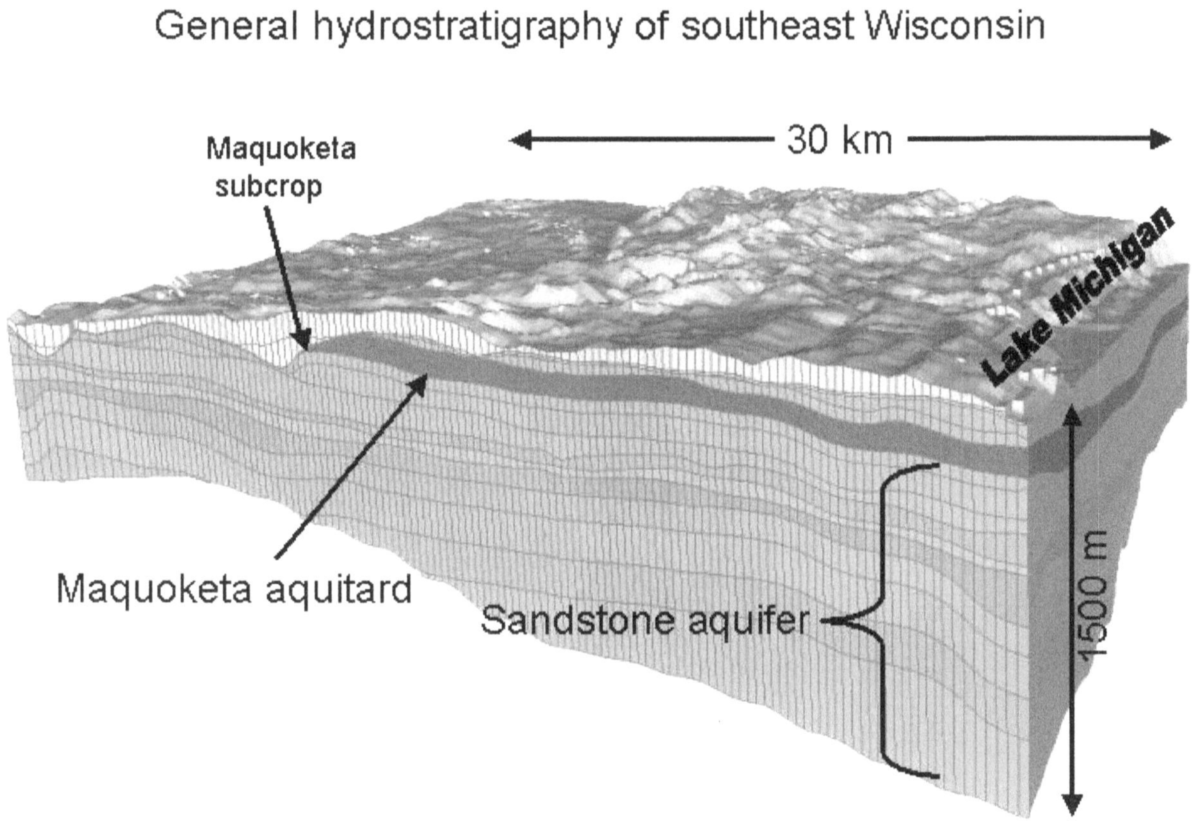

Source: Feinstein et al. 2004

Figure 2.9 Schematic view of the Maquoketa Formation in southeast Wisconsin

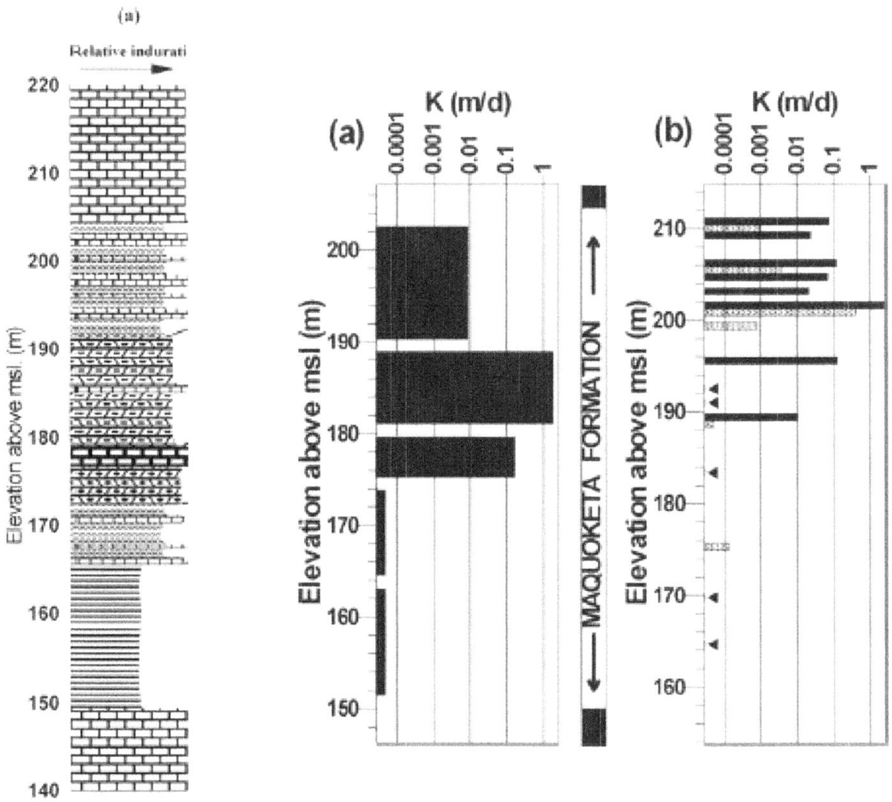

Source: Eaton 2002

Figure 2.10 Profiles of hydrostratigraphy of the Maquoketa Formation in Eastern Wisconsin. Left: lithology. Right: horizontal hydraulic conductivity distribution based on packer testing of vertical boreholes at two sites

CHAPTER 3

GROUNDWATER FLOW THROUGH AQUITARDS

FLOW ACROSS AQUITARDS IN REGIONAL GROUNDWATER FLOW SYSTEMS

In the broadest generalization of natural groundwater flow systems composed of both aquifers and aquitards, flow across aquitards is primarily vertical and flow in aquifers is horizontal (Figure 1.1). Conceptually, aquitards are integral parts of regional flow systems, in which the presence of these low hydraulic conductivity formations commonly increases the complexity of flowpaths (Freeze and Witherspoon 1967). Zijl and Nawalany (1993) demonstrated the propensity for vertical flow in aquitards using a two-dimensional numerical model to simulate groundwater flow in geologically simple systems. Figure 3.1(a) shows the flow domain with boundary conditions and 3.1(b) shows the flow lines (e.g., trajectories of water particles for the case of a homogeneous aquifer). The flow system is symmetrical with short shallower flow paths in the center of the domain, and deep, long flow paths on the sides and bottom. The insertion of the single, horizontal aquitard into the system separates the flow domain into three zones: upper aquifer, aquitard and lower aquifer. Because the sides and bottom of the flow domain are impervious, strong vertical hydraulic gradients form across the aquitard to maintain continuity of flow. Beneath the recharge area, the flow lines are primarily vertical, directed downward in the aquitard and upward beneath the discharge area. Much of the groundwater circulating in the system is restricted to the upper aquifer. The age of groundwater in the lower aquifer is much greater than in the upper aquifer because of slow travel through the aquitard. Figure 3.1(d) shows a well pumping gently in the lower aquifer extracting only older groundwater. However, a strongly pumped well draws flow lines down through the aquitard into the well, in which case the entire surface of the flow system would become a recharge area.

Aquitards may have gaps (e.g., windows) where the low hydraulic conductivity (K) material is absent (Figure 1.2(c)). In the vicinity of the window, the groundwater zones above and below the gap can be weakly or strongly connected, depending on the site-specific circumstances. Figure 3.2(a) is an example of a single gap in a horizontal aquitard. The groundwater in the lower aquifer is separated from the water in the upper aquifer under natural flow conditions, however, operation of a pumping well in the lower aquifer can cause strong changes to the flow system. Figure 3.2 shows two scenarios for a pumping well situated in the lower aquifer near one of the gaps in the central part of the flow system. In the initial condition (Figure 3.2(b)), the well is pumped only gently and there is minimal influence on the flow patterns. In the second condition (Figure 3.2(c)), the pumping well receives substantial flow from the upper aquifer because of the strong induced drawdown gradients and proximity of the gap to the well. Thus, if the shallow aquifer is contaminated in the vicinity of the window, there is strong potential for contamination to influence water quality of the pumping well. In the assessment of aquitard integrity, determination of the presence and location of windows and their influence on groundwater flow paths through aquitards is an important endeavour.

Freeze and Witherspoon (1967) conducted the first analysis of groundwater flow paths in steady-state systems comprising aquifers, aquitards and irregular topography with these factors arranged with some complexity. Using a two-dimensional numerical model, they showed that the presence of geologic layers offering only small differences in hydraulic conductivity (e.g., K contrasts of only 10 or 100) caused strong effects on the pattern of groundwater flow

(Figure 3.3). Groundwater flow in the higher K units is generally horizontal and many of the flowlines are generally vertical or nearly so in the lower K units. For multi-aquitard systems, this tendency for flow to be primarily vertical in the aquitards is conceptualized in Figure 1.1. Tóth (1995) added the concept of groundwater age zones to the flow diagram. In Figure 1.1, the aquitards are continuous across the flow domain so that all groundwater below the shallowest aquitard has passed slowly through the aquitard. Therefore, groundwater in the deeper aquifers is old compared to groundwater in the surficial aquifer. This illustrates a characteristic situation of groundwater flow systems – the occurrence of very old groundwater in aquifers located beneath unfractured aquitards with lateral continuity.

In the aquitard context, we refer to flow systems as "steady-state" or "transient" depending on whether or not there is equilibrium within the aquitard with respect to present-day hydrogic conditions. The time scales of flow (Alley et al. 2002) through aquitards may range from less than a year to millennia or longer according to their hydraulic properties, hydraulic gradients and aquitard thicknesses and assuming porous media flow. In some cases, regional groundwater flow systems are not in equilibrium with the existing climatic conditions (e.g., recharge) because the hydraulic equilibrium times for aquitards are slower than the rate of climatic change. In some cases, the rates of withdrawal from aquifers change rapidly and the pore-water pressure distribution in the associated aquitard takes much longer to adjust (e.g., Husain, Cherry, and Frape 2004). Although very old groundwater commonly exists in aquifers positioned beneath extensive aquitards, once the aquifer is strongly pumped, the flow line positions and groundwater velocity can change, in some cases resulting in much younger water arriving in the aquifers. This may markedly increase the potential for wells to produce contaminated water.

At the regional scale, groundwater flow is three-dimensional, and highly influenced by the geometry and hydraulic properties of geological formations (Eaton 2002, Fogg 1990, Macfarlane et al. 1994). Figure 3.4 is a conceptual diagram of regional flow in a typically complex subsurface environment. From the regional-scale perspective of a system with multiple pumping wells, this diagram illustrates flow is neither strictly horizontal in aquifers nor strictly vertical in aquitards. The aquitards deflect and direct flow, and control boundaries between flow systems at local, intermediate and regional scales.

Complex groundwater flow paths can occur at the edges of regional aquitards. For example, in southeastern Minnesota, the Decorah-Platteville-Glenwood aquitard confines the St. Peter-Prairie du Chien-Jordan aquifer, a major source of municipal water supply. Where this aquitard appears in a bluff at land surface, model results indicate the diversion of flowpaths around the subcrop causes greatly increased recharge to the underlying aquifer (Lindgren 2001). Preferential flow may also be important in other areas of similar geometry where the edges of aquitards in the deep subsurface are less amenable to study. Regional flow modeling results for southeastern Wisconsin suggest that a disproportionate number of flowpaths to pumped wells in the deep confined bedrock aquifer originate west of the subcrop of the major regional aquitard, the Maquoketa Formation (Feinstein et al. 2004).

Pumping wells installed in aquifers (Figure 3.2) can alter flow paths. The pumping can reverse hydraulic gradients and flow directions so groundwater previously discharged to surface waters now flows toward wells. This reversal of flow directions can decrease aquitard integrity. Fractures or windows in the aquitard are now more likely to allow contamination to move into the lower aquifer. Depending on the time scales in different parts of the system, flow on a local scale may reach a steady state after some time has passed, while flow on a regional scale remains transient. For example, in Figure 3.4, the discharge of wells in shallow aquifers may be completely accounted for by induced recharge from nearby rivers (well B) or through thin aquitards (well F). In such cases, drawdown in these pumping wells is no longer increasing for a constant pumping rate. In other cases, discharge from deeper wells in larger flow systems (wells E, H) comes from regional recharge.

HYDRAULIC HEAD DISTRIBUTION

For contaminants in the dissolved and particulate categories, groundwater flow causes transport along groundwater flow lines. In the case of DNAPL, groundwater flow enhances downward movement when the hydraulic gradient is downward; downward DNAPL movement is diminished or the direction may be reversed when the hydraulic gradient is strongly upward. Determining the propensity for groundwater flow to influence contaminant migration across or within aquitards requires determining the hydraulic gradient in the aquitard in addition to obtaining other information about aquitard properties, such as hydraulic conductivity and porosity. As described in Section 3-1, in non-indurated hydrogeologic domains composed of flat-lying aquifers and aquitards, the hydraulic gradient within aquitards is typically vertical or near-vertical and the hydraulic gradient within the aquifers is primarily horizontal. However, the presence of a strong vertical gradient across the aquitard does not preclude situations where stratification of hydraulic conductivity distribution within the aquitard causes local zones of horizontal flow.

The direction of groundwater flow across a buried aquitard is commonly determined by measuring the head differential across the aquitard (Figure 3.5a). For a surficial aquitard, the steady state gradient across the aquitard is determined by comparison of the water-table elevation and the head in the aquifer below the aquitard (Figure 3.5b). In the situation in Figure 3.6, the direction of the gradient (up or down) and the average gradient represented by the straight line in the hydraulic head profile is obtained using only two piezometers or monitoring wells. For the buried aquitard case (Figure 3.5a), these two measurements are generally made external to the aquitard; for the surficial aquitard (Figure 3.5b), one is external (i.e., in the underlying aquifer) and the other internal (at the water table). The straight-line representation of the head profile across the aquitards is appropriate when aquitard homogeneity is assumed (Figure 3.6). This approach using measurements of head external to the aquitard is common and accomplishes the task of determining whether groundwater flows upward or downward across the aquitard and the overall hydraulic gradient across the aquitard. However, much more information relevant to aquitard integrity is obtained when head measurements are made inside the aquitard (e.g., a vertical profile of head versus depth across the aquitard). Figure 3.7 shows a schematic head profile through an aquitard obtained from a piezometer nest or a multilevel monitoring system. The hydraulic gradient measured within the aquitard at any particular elevation is generally different from the average gradient across the aquitard obtained from external measurements. The internal variability in the vertical gradient distribution in the aquitard is caused by variability (i.e., heterogeneity) of the hydraulic conductivity in the vertical direction (K_v).

Freeze and Cherry (1979, p. 33) show a relation between layered heterogeneity, bulk hydraulic conductivity and Darcy flow. Consider the layered aquitard in Figure 3.8. Each layer is homogeneous and isotropic with hydraulic conductivity values of K_1, K_2 … K_n but the system as a whole acts like a single, anisotropy layer. For flow continuity under steady state conditions, the Darcy flux (q) (i.e., specific discharge) for each layer is the same entering as leaving, and must be the same throughout each layer and from layer to layer. The head loss across the entire aquitard is the sum of the losses across each layer: $\Delta h = \Delta h_1 + \Delta h_2 + ... + \Delta h_n$ and from Darcy's Law:

$$q = \frac{K_1 \Delta h_1}{\ell_1} = \frac{K_2 \Delta h_2}{\ell_2} = \ldots \frac{K_n \Delta h_n}{\ell_n} = K_z \frac{\Delta h}{\ell_n} \tag{3.1}$$

And therefore:

$$\Delta h = \frac{\ell_n}{K_z} = q \left[\frac{\ell_1}{K_1} + \frac{\ell_2}{K_2} + \frac{\ell_3}{K_3} \ldots \right] \tag{3.2}$$

where q = Darcy flux
K = hydraulic conductivity
h = hydraulic head
ℓ = layer thickness

This analysis indicates the largest head loss occurs across the layer with the smallest K_v when thickness does not vary from layer to layer. For example, in an aquitard with two layers, if one layer has a K_v 100 times greater than the other layer, essentially all of the head loss (> 99%) occurs across the lower K_v layer. Figure 3.9(a) illustrates this situation in which the upper aquitard layer is composed of silt and the lower layer is clay. In Figure 3.9(b), the entire aquitard layer is clay, but has vertical fractures of different lengths and the lower zone of the aquitard is unfractured. In both cases, the large difference in head loss between the upper and lower aquifer zones is indicative of a major difference in the internal distribution of vertical K in the aquitard. However, the head profile alone is not sufficient for identification of the cause of the large K_v difference between the upper and lower zones. Core sample logs or bore hole geophysics indicating the geologic nature of the two zones are needed to complement the head profile. For the situation in Figure 3.9(b), if the composition of the aquitard was known to be the same geologic material throughout (e.g., same texture / lithology), the head profiles would be strong evidence indicating the presence of vertical fractures to considerable depth.

Figures 3.10 and 3.11 show two field examples in which multilevel monitoring of hydraulic head was conducted across aquitards: one composed of a silty clay and the other of shale (indurated rock). For each aquitard, these examples show nearly all of the head loss across the aquitard occurring in a thin zone at the aquitard bottom. In the case in Figure 3.10 (from Meldrum 1999), the entire aquitard (8 m thick) is composed of clayey silt of glaciolacustrine origin with several thin horizontal sand seams. Nearly all of the head loss measured between two of the ports on the multilevel system occurs at the bottom of the aquitard. These two ports are only 0.4 m apart and the head loss across this distance is 3 m. Therefore, the hydraulic gradient at the bottom of this aquitard measured by the two ports is 7.5. The actual maximum gradient at the bottom of the aquitard may be larger because the low K zone between the two ports may be thinner than the distance between the ports. These figures illustrate a condition not unusual in clayey aquitards, which is the vertical hydraulic gradient greater than 1. Morrison, Parker, and Cherry (1998) showed tetrachloroethylene (PCE) contamination occurs in the aquitard shown in Figure 3.10 due to the experimental release, described by Brewster et al. (1995), of PCE DNAPL into the surficial sand aquifer. The DNAPL penetrated deep into the aquitard via vertical fractures but did not go all the way through to the underlying aquifer. The lack of penetration through the bottom part of the aquifer, where the head profile indicates much lower K_v, is probably due to less fractures or absence of fractures in this basal zone.

In the shale case in Figure 3.11 (from Eaton and Bradbury 2003), the head profile also shows an extremely large hydraulic gradient across a zone at the bottom of a regional shale aquitard in Wisconsin. The gamma log indicates the shale unit is 60 m thick but the zone of large head loss is only 4 m thick and therefore the vertical hydraulic gradient across this zone is 10. The two field examples in Figures 3.10 and 3.11 show a hydrogeologic unit designated as an aquitard based on stratigraphic or lithologic considerations alone can be much thicker than the zone within the aquitard that actually causes nearly all of the head loss and low groundwater flux across the aquitard. Thus, the most direct and reliable means for identifying the zones with lowest K_v within an aquitard is measurement of the hydraulic head profile. Hydraulic conductivity tests in boreholes using straddle packers provide values for horizontal K that are generally not a reliable representation of vertical K. Laboratory measurements of K_v using borehole core samples do not necessarily indicate the actual bulk vertical K distribution because they generally do not detect fractures. Other lines of evidence need to be assessed to deduce whether or not the K_v values from lab tests on core samples represent the bulk aquitard K_v.

The need for a large number of hydraulic head measurements to produce profiles across aquitards provides impetus for use of depth-discrete multilevel monitoring systems with borehole seals between many measurement ports (Figure 3.7(b)). Four general types of multilevel systems suitable for measuring head profiles in aquitards and in aquifer-aquitard systems are available in the commercial market. These four systems, described by Einarson (2005), are very different in design but all provide head values and water samples when used in the appropriate hydrogeologic condition. Information about these systems is available from: Flexible Liner Underground Technologies (www.flut.com), Solinst Canada (www.solinst.com) and Westbay Instuments (www.westbay.com).

One of the four systems (Waterloo-Solinst) has recently been adapted specifically for installation in holes created by rotosonic drilling. This type of drilling equipment has become available from contractors in nearly all parts of North America and in investigations of aquitard integrity, it offers some advantages over conventional methods: i) nearly all types of non-lithified aquitards can be drilled through without encountering refusal, ii) continuous cores are produced and iii), relatively large drill casing (5.5-7 inch ID) is used, which provides a reliable means for installing multilevel systems in just about any hydrogeologic circumstances. Parker, Cherry, and Swanson (N.d.) describe the adapted version of the Waterloo-Solinst multilevel system, which they used in two aquitard investigations. This adapted system provides a maximum of 15 monitoring ports in a single hole with secure seals between each port. Large numbers of monitoring ports are advantageous for determining the depth intervals where the lowest vertical hydraulic conductivity occurs, as illustrated in Figures 3.10 and 3.11. These low conductivity zones are the intervals within an aquitard where most or nearly all of the protective capacity exists. These critical zones within an aquitard can be very thin. Large numbers of ports in each multilevel system provide closest vertical spacing between ports, which provides best opportunity to determine presence and thickness of critical zones. Rotosonic drilling also provides excellent opportunity for installing other types of commercial multilevel systems with secure seals between ports.

The information presented above for hydraulic head in aquitards considered only steady-state conditions, however important information can also be obtained from temporal measurements of hydraulic head in aquitards over short- and long-time periods. For example, several studies have used long-term head variations in aquitards as evidence for extremely low vertical hydraulic conductivity in aquitards (Eaton 2002, Eaton and Bradbury 2003, Husain et al.

1998). In these studies, the very low K_v of the aquitards allowed disequilibrium to persist for many years or decades relative to the head in pumped aquifers. This approach to the determination of the bulk K_v for the aquitard is similar to that used for pumping test analysis, except the aquifer drawdown is caused by groundwater resources extraction over a long time period rather than the short pumping period used in a pumping test.

DETERMINATION OF HYDRAULIC CONDUCTIVITY OF AQUITARDS

Measurements of aquitard hydraulic conductivity along with hydraulic head measurements provide the basis for calculating groundwater flux through aquitards. These measurements also provide insight into the nature of the groundwater flow paths in the aquitard (e.g., inter-granular flow paths versus fracture flow paths). Groundwater velocity calculations also require determinations of porosity. Haefner (2000), Neuzil (1986) and van der Kamp (2001) provide reviews of aquitard hydraulic conductivity and the methods used for its determination. There are two general categories of methods for determining the hydraulic conductivity of aquitards: external and internal. External measurements include pumping tests in which an aquifer above or below the aquitard is pumped or stressed while monitoring the head (e.g., drawdown) in the pumped aquifer and, in some cases in a second aquifer above or below the aquitard. This is the traditional pumping test approach in which the bulk K_v of the aquitard is inferred from the head response data external to the aquitard. Figure 3.12 shows schematically the type of drawdown data produced by such tests. The apparent leakage effect identified on the drawdown graphs may not be caused entirely by aquitard leakage because heterogeneities and boundaries can provide similar effects in the aquifer; the apparent K_v can be much different than the actual bulk K_v representative of the aquitard. Batu (1998) provides a review of methods used to calculate K_v values from aquifer pumping tests.

Pumping tests are conventional in the sense they have been in common use for many decades, however the common goal of pumping tests is to determine aquifer properties and aquifer yield, and knowledge of aquitard properties is usually not a primary pursuit. In particular, for purposes of aquifer yield, knowledge of the specific nature of the groundwater flow paths through the aquitard is not needed as long as the quantity of flow across the aquitard is predicted. The external approach to determining aquitard K_v has severe limitations when the goal is assessment of aquitard integrity.

Pumping tests conducted for the specific purpose of aquitard integrity assessment require measurement of hydraulic head within the aquitard and in the pumped aquifer. Wolff (1970) was the first to report a pumping test with head response measurements in both the aquifer and aquitard. Wolff used a one-dimensional mathematical model to interpret the results. In this case, the aquitard response agreed closely with the predictions based on laboratory measurements of aquitard properties using core samples, and therefore the test provided no indication of fractures or other preferential pathways. An analytical, two-dimensional "ratio method" using type curves for aquitard response was developed by Neuman and Witherspoon (1972) for aquifer pumping tasks to investigate aquitards. This analytical method was extended by Rowe and Nadarajah (1993) to effects of piezometer time lag and the vertical length of the piezometer monitoring interval. Numerical methods can also be used, which allow for more complex aquifer-aquitard configurations, such as pumping from multiple wells (Neuman and Gardner, 1989), for layered aquitards (Rowe, Quigley, and Booker 1995), or for changes of permeability and storage coefficient within the aquitard as it undergoes consolidation (Rudolph and Frind 1991). Grisak and Cherry (1975) describe a pumping test in a surficial clayey aquitard in which both aquitard internal and external head measurements were made in many piezometers. They concluded that the much larger aquitard K_v determined from the pumping test was caused by vertical fractures penetrating from the water table down through the aquitard, which provided hydraulic connection with the underlying aquifer. Some of the piezometers in the aquitard responded quickly to aquitard head drawdown and others did not respond, indicating some piezometers situated on or near fractures and others in the matrix blocks between fractures, as illustrated in Figure 3.13. Piezometers in large matrix blocks respond at rates governed by K_v values measured on core samples.

At the other end of the spatial scale for aquitard hydraulic conductivity are measurements determined in the laboratory on core samples. For clayey aquitards, K_v and compressibility (storativity) values are determined using standard triaxial or consolidometer devices used in the geotechnical field (American Society of Civil Engineers 1960). For rock samples, triaxial and other types of geomechanics test equipment are used. These laboratory measurements on core samples from vertical holes produce values of K_v and storativity for the aquitard matrix. The core specimens tested are typically only a few centimeters in diameter and 5-10 centimeters long. For unfractured aquitards comprised of non-indurated clayey deposits, K_v values determined using such small core samples may provide a representative K_v value for the aquitard, however these core-derived K_v values need to be assessed for representativeness by comparison to field-derived values. In the context of the aquitard at the field scale, K_v values determined from core samples may provide minimum values because core samples have a low probability of containing open, vertical fractures or other preferential pathways such as root holes. Vargas and Ortega-Guerrero (2004) used comparison of slug tests in piezometers and laboratory measurements of hydraulic conductivity as the primary evidence supporting the conclusion clayey aquitards in the Mexico City area have fractures to a depth of 40m. However, for some aquitards the core test K_v values are similar to the values determined from field hydraulic tests providing no indication of preferential flow paths (e.g., Shaw and Hendry 1998).

Many investigators (Bradbury and Muldoon 1990, Clauser 1992, Schulze-Makuch et al. 1999, Schulze-Makuch and Cherkauer 1998) report the measured hydraulic conductivity of a hydrogeologic unit increases with measurement scale, sometimes by several orders of magnitude. However, whether this apparent "scale effect" is a fundamental property of porous materials or simply a manifestation of bias inherent in the various measurement methods is problematic (Zlotnik et al. 2000). Hsieh (1998), in a recent review of this phenomenon, suggests the scale effect might be more reasonably termed "sampling bias." However, both Hsieh and Zlotnik et al. examined this topic in the context of broader context of permeable geologic media without specific consideration of aquitards. The scale-effect phenomenon is particularly important in aquitards because there has been a tendency to extrapolate measurements made at the core or piezometer scale to site and regional scales. Many investigators (Bradbury and Muldoon 1990, Grisak and Cherry 1975, Grisak et al. 1976) present field data for sites where laboratory measurements of hydraulic conductivity in till are significantly lower than field measurements of the same materials. At waste disposal sites in eastern Wisconsin, Bradbury and Muldoon (1990) attributed the discrepancy to a combination of directional bias and undersampling of the more-conductive heterogeneities. Vertical fractures or stratigraphic windows are important, but the results of laboratory measurements on core samples rarely reflect these features. Therefore, the challenge is determining whether or not such features are present using laboratory core measurements along with other types from larger scale.

In addition to the pumping test methods described above, five other methods exist for measuring the hydraulic conductivity of aquitards: 1) slug tests in standpipe piezometers or wells, (e.g., Butler et al. 1994), 2) porewater pressure changes in the interior of aquitards propagated downward from natural fluctuations in the shallow water levels caused by precipitation events, which is analogous to a pumping test (e.g., Davis 1972; Keller, van der Kamp, and Cherry 1989; Boldt-Leppin and Hendry 2003), 3) long term flow into a large augered cavity (Keller, van der Kamp, and Cherry 1989), 4) aquitard response to loading or unloading due to engineering construction (van der Kamp and Maathuis 1985), and 5) calibration of two or three-dimensional numerical models for groundwater flow based on hydrostratigraphic information and field measurements of hydraulic head at strategic locations (e.g., Grisak and Cherry 1975, Krohelski et al. 2000, Feinstein et al. 2004). Slug tests have limited usefulness for determination of K_v because, in vertical holes, slug response commonly depends primarily on the horizontal component of the hydraulic conductivity. However, slug tests can indicate the presence of permeable zones, providing valuable insight concerning the internal nature of the aquitard. If the slug test is conducted in a zone in a clay aquitard where other lines of evidence (e.g., core inspection) indicate no granualar material (e.g., sand seams), and if the slug test result is significantly larger than results from laboratory measurements on cores, the presence of fractures may be the most plausible explanation. Or, if the slug test provides higher K values only due to coarser grained layers in the aquitard, these values may have no relevance to determination of the K_V.

With regard to the determination of the permeability of non-indurated aquitards, van der Kamp (2001) states:

> "with judicious selection of the most suitable methods for a particular site, good test design, careful instrumentation, and respect for the underlying assumptions, reliable determinations of aquitard permeability can be obtained".

Keller, van der Kamp, and Cherry (1989) present results of hydraulic conductivity measurements using the first three of the five methods listed above, with the results in close agreement because the investigated aquitard is composed of nearly homogeneous unfractured silty clay glacial till except at shallow depth. Slug test results are controlled by the hydraulic conductivity of the materials very close to the well screen or piezometer tip, including the sand pack if present. Effects of drilling on the borehole wall can be important. For example, drilling, particularly auger drilling, may create a smear zone at the borehole wall with a much lower hydraulic conductivity than the natural geologic media away from the wall. D'Astous et al. (1989) describe a method for minimizing shear effects in clayey aquitards. Li and Jaio (2001a, 2001b) evaluated the effect of tidal fluctuations on hydraulic properties and leakage in semi-confined marine aquitards.

Relative to what is known about aquitards based on field investigation, commonly-used numerical models of groundwater flow and transport simplify or idealize aquitards. For example, the USGS MODFLOW code (McDonald and Harbaugh 1988) provides an option for representing aquitards as quasi-confining beds having negligible thickness and no storage or horizontal flow. This simplification saves computer time and storage, and is adequate for simulations unconcerned with contaminant transport or lateral components of flow in the aquitard, but is inappropriate for many contaminant transport problems. MODFLOW can also simulate aquitards as distinct hydrostratigraphic model layers having storage, horizontal flow, and areally variable properties. For each model layer the code calculates vertical flux based on the vertical hydraulic gradient across the aquitard layer and a parameter called VCONT that represents the harmonic mean vertical leakance across the layer. Regional flow models that represent aquitards using one or two hydrostratigraphic model layers cannot simulate the complexity in vertical head gradients often found, and therefore many locally overestimate or underestimate their effective vertical hydraulic conductivity.

Indirect methods such as environmental isotopes and hydrogeochemistry have been useful for estimating hydrologic properties for unlithified aquitards and for obtaining other insights relevant to aquitard characteristics (e.g., Desaulniers, Cherry, and Fritz 1981; Desaulniers et al. 1986; Remenda 1993; Remenda, Cherry, and Edwards 1994; Remenda, van der Kamp, and Cherry 1996). These methods are discussed further in Section 4.8.

Hydrogeologic Properties of Aquitards Reported in the Literature

One of the better-known regional bedrock aquitards for which considerable hydrologic data have been published is the Pierre Shale (Neuzil 1986; Neuzil 1988; Neuzil 1993; Neuzil, Bredehoeft, and Wolff 1984) in the U.S. northern Great Plains. Although lithologically variable (Schultz et al. 1980), this aquitard appears hydrologically massive and homogeneous where it has been studied in South Dakota (Neuzil 1993). A discrepancy exists between in-situ values and estimates of inverse-model-derived regional scale hydraulic conductivity (Bredehoeft, Neuzil, and Milly 1983). The model-derived values are three or four orders of magnitude greater. Widely spaced vertical fractures are hypothesized to explain this discrepancy (Neuzil, Bredehoeft, and Wolff 1984). Indications of preferential flow based on coincidence of lineaments and hydrochemical or geothermal anomalies (Kolm and Peter 1984) can provide insight relevant to possibilities for possible contaminant transport in indurated aquitards.

Hydrogeologic information has been reported for other bedrock aquitards. Solute transport through Cretaceous clays with similarities to the Pierre Shale has been investigated in Canada (Cherry 1989, Mase et al. 1987, Shaw and Hendry 1998). Fractured aquitards formed in chalk have been studied in Israel (Rophe et al. 1992) and in Texas (Dutton et al. 1994, Wang and Myer 1994). Hydrogeologic characteristics of shale aquitards have been characterized in Wisconsin (Eaton 2002; Eaton, Anderson, and Bradbury 2001; Eaton et al. 2000), Illinois (Graese et al. 1988) and Ontario (Weaver 1994). European studies supporting site selection for nuclear waste repositories are focused on shale and argillaceous marl in the Alps (Gautschi 2001, Mazurek et al. 1998, Thury and Bossart 1999), mudstones in Great Britain (Sen and Abbott 1991), and shale in France (Boisson et al. 2001, Bonin 1998), Marty, Dewonck, and France-Lanord 2003). The impact of mudstone aquitards on regional flow has been studied in Kansas (Macfarlane et al. 1994). Hydraulic characteristics of mudstones are known from the Newark basin in the northeastern U.S. (Maguire 1998, Michalski and Britton 1997). Nativ, Halleran, and Hunley (1997) and Schreiber, Moline, and Bahr (1999) reported contaminant transport in, and hydraulic properties of, a highly weathered, deformed shale in South Carolina.

Hydrogeologic information on bedrock aquitards can be obtained through indirect studies using environmental isotopes (Dutton and Simpkins 1989, Hendry and Wassenaar 1997, Skilliter 2000, Stueber and Walter 1994, Stueber et al. 1993) and tracers (Bishop et al. 1993a). Hydrochemistry has elucidated fracture flowpaths in chalk (Adar and Nativ 2000, Nativ et al. 1995, Nativ and Nissim 1992). When no hydrologic information exists for a given bedrock aquitard, there may be value in drawing analogies to hydrologic studies of similar bedrock types elsewhere. Information is available on fine-grained, low conductivity rocks in regional hydrogeologic frameworks, for example the Chattanooga and Ohio Shales in the Appalachian basin (Seaber, Brahana, and Hollyday 1998), the Green River Formation on the Colorado Plateau

(Taylor and Hood 1988) or the Permian evaporites underlying the Central Great Plains (Jorgensen et al. 1988). Stratigraphic and petroleum studies (Hearn 1996, Hornung and Aigner 1999, Macfarlane et al. 1994, Weaver 1994) or sedimentological and engineering studies (Kepferle, Potter, and Pryor 1981; Russell and Harman 1985; Wang and Myer 1994) also provide information on aquitards. Table 3.1 compiles hydrogeologic data from studies in different unlithified aquitards. Table 3.2 compiles hydrogeologic data from studies of bedrock aquitards.

Table 3.1 Selected hydrogeologic data on unlithified aquitards

Formation Name, Location	Lithology	Test Method	Hydraulic Conductivity K	Storage, dispersivity, or velocity values	Porosity, fracture apertures, densities	Reference
Ringe till, (Ringe, Funen, Denmark)	Massive, bioturbated weathered and unweathered, vertically fractured clayey till	Column tests (0.1m?); Piezometers (1m); Infiltration tests (10s m)	2.5 to 3×10^{-6} m/s; 9.5×10^{-7} to 9.7×10^{-6} m/s; 1×10^{-5} to 7×10^{-5} m/s (weath), 5×10^{-8} to 1.95×10^{-5} m/s (less fract. or unweath)	N/A	Matrix total porosity 0.25-0.35	Jorgensen et al. 1988, Nilsson et al. 2001
Ringe till, (Ringe, Funen, Denmark)	Massive, bioturbated weathered and unweathered, vertically fractured clayey till	Darcy Law calc. on flow to piezometers (1m); Discrete fracture (**DF**) model fitting of tracer test breakthroughs	0-2.5 m depth: $K=8 \times 10^{-6}$ to 6×10^{-5} m/s; 0-4m depth: $K=7 \times 10^{-6}$ to 3×10^{-5} m/s	Flow velocities in fractures 107 to 349 m/d	Desicc./tect. fract. density 2-3.3 m^{-1} to 2.4m depth; hydr. apert. 110-158 μm to 2.5 m depth, 65-127 μm to 4 m depth	Sidle et al. 1998
St. Joseph till, (Sarnia, Ontario, Canada)	Massive weathered and unweathered, vertically fractured clayey till	Lab (**L**) oedometer (0.01 m); piezometers, seepage collectors (1m); trench infiltr (10s m); cubic law fracture model	**L** 1.2 to 2.1×10^{-10} m/s (weath), 1.5 to 3.1×10^{-10} m/s (unweath); piezo (weath) 10^{-10} to 3×10^{-7} m/s; seep. 5×10^{-10} to 1×10^{-6} m/s; trench 1.6 to 2.6×10^{-7} m/s	Lab storage: 0.4 to 1.0×10^{-3} m^{-1} (weath), 1.0 to 5.0×10^{-3} m^{-1} (unweath); most calculated velocities 0-5 m/d, up to 24 m/d.	Fracture porosity 10^{-3} to 5×10^{-5}; fract. densities 100 m^{-1} to <1 m^{-1} decr. with depth; hydr. Apert. most <21 μm, up to 43 μm;	McKay et al. 1993
St. Joseph till, (Sarnia, Ontario, Canada)	Massive weathered, vertically fractured clayey till	Simulation of solute and colloid tracer breakthrough, in weathered zone (<3m depth), with discrete fracture (**DF**) and **EPM** models	Solute in piezo (**DF**) 1.2×10^{-8} to 2.7×10^{-7} m/s; Solute via seepage (**DF**) 3.4 to 4.2×10^{-7} m/s; Coll. via seepage (**DF**) 5×10^{-8} to $>2 \times 10^{-7}$ m/s; **EPM** K_b 6×10^{-8} to 3×10^{-7} m/s	Solute seepage D_L 0 to 0.1 m, velocity 10-17 m/d; Colloid seepage velocity 2 to >5 m/d; Solute in piezometers D_L 0 to 0.2 m, velocity 2.3-13.7 m/d	Total matrix porosity 0.32; Fracture porosity 10^{-5} to 10^{-3}; Apertures from solute seepage 19 to 37 μm;	McKay, Cherry, and Gillham 1993

(continued)

Table 3.1 (Continued)

Formation Name, Location	Lithology	Test Method	Hydraulic Conductivity K	Storage, dispersivity, or velocity values	Porosity, fracture apertures, densities	Reference
Various studies, Sweden and Norway	Structured but apparently unfractured sandy-silty tills	Field pump (10s m), infiltr pit, ring infiltr (1m); lab permeameter (0.01 m)	Field $2x10^{-8}$ to $5x10^{-4}$ m/s; Lab $5x10^{-9}$ to $5x10^{-5}$ m/s	N/A	Total porosity 0.18-0.48; eff. poros 0.02-0.33; charact. pores 30-95 µm	Lind and Lundin 1990 and references therein
Various studies in Denmark	Unweathered and unweathered fractured and unfractured clayey till	Trench drainage, slug tests, lab triax testing; field (**F**) pumping, leakage est., tritium distrib. (1-10 m); Lab (**L**) perm., triax, oedometer (0.01 m)	Trench $>10^{-7}$ m/s, slug tests in weath., fract. $7x10^{-8}$ to $3x10^{-6}$ m/s, triax unweath. 3-$6x10^{-10}$ m/s, triax weath. 1-$2x10^{-9}$ m/s; **F** Kv $5x10^{-10}$ to $2x10^{-6}$ m/s; **L** $2x10^{-11}$ to $5x10^{-9}$ m/s	N/A	Horizontal and vertical fracture density 1-3.3 m^{-1} to 6 m depth; Eff. porosity 0.01 (fractured?) to 0.30 (unfractured ?)	Fredericia 1990 and references therein
Wisconsin-age weathered till, Iowa, US	Weathered and fractured sandy loam till	Const.flowrate (**CF**) and const. drawd. (**CD**) pumping tests (250 m^3); bail (**B**) tests (1 m^3).	**CF** $3.7x10^{-6}$ to $7.5x10^{-6}$ m/s; **CD** $3.3x10^{-6}$ to $9.7x10^{-6}$ m/s; **B** $5.2x10^{-8}$ to $4.7x10^{-6}$ m/s	Specific yield **CF** 0.015-0.065; **CD** 0.020-0.089; est. lateral porewater velocity 53 m/yr	Total volumetric porosity 0.32; effective porosity 0.032-0.064.	Jones, Lemar, and Tsai 1992
Oak Creek Formation, southeastern Wisconsin, US	Weathered and fractured, and unweathered clayey silt till interbedded with silty sand and clayey silt	Field (**F**) slug tests (1 m^3) and lab (**L**) triaxial cell tests (0.1 m^3)	**F**: till $1x10^{-10}$ to $4.3x10^{-6}$ m/s, cly silt to slty sand $2x10^{-9}$ to $3x10^{-5}$ m/s; **L**: till $1x10^{-11}$ to $2x10^{-6}$ m/s, cly silt to slty sand $5x10^{-11}$ to $8.4x10^{-6}$ m/s	Avg. field linear velocity 0.011 to 0.411 m/yr	Bulk porosity 0.21-0.49; Effective fracture spacing 4 m.	Simpkins 1989

(continued)

Table 3.1 (Continued)						
Formation Name, Location	**Lithology**	**Test Method**	**Hydraulic Conductivity K**	**Storage, dispersivity, or velocity values**	**Porosity, fracture apertures, densities**	**Reference**
Fluvial, marine deposits and peat, Schelluinen, Netherlands	Flood basin clays, crevasse-splay sand to silty clays, organic peaty clay and peat	Core permeametry (10^{-1} to 10^{0}m); Block upscaling (10^{1} to 10^{2} m); regional modeling (10^{3} to 10^{5} m)	Mean K's $1.5x10^{-8}$ to $2x10^{-5}$ m/s; Block mean K's: $2.8x10^{-8}$ m/s to $3.6x10^{-5}$ m/s; effective Kv* $5x10^{-8}$ to 10^{-7} m/s	Max velocities from particle travel time distributions 0.03 to 0.04 m/d across 8.5 m thickness	Effective porosity (core data and capillary model) 0.08-0.34	Bierkens 1996 *Effective Kv est. from hydraulic resist, head, thickness, flux
Livingston hazardous waste landfill, Louisiana Gulf Coast, USA	Paleoweathered, jointed, fluvial, deltaic and marine silts and clays	Laboratory core Kv (0.01 m); vertical Kv from water balance (WB) model of waste cells (1000s m^2)	Clays 10^{-11} to 10^{-8} m/s; silty, sandy clays $5x10^{-11}$ to $3x10^{-8}$ m/s; slickens. 10^{-10} to $2x10^{-7}$ m/s; org. $8x10^{-10}$ to $3x10^{-7}$ m/s; silt 10^{-10} to 10^{-6} m/s; sand $8x10^{-9}$ to $5x10^{-7}$ m/s; WB 10^{-7} m/s	N/A	Estimated sediment porosity 0.25	Hanor 1993

Table 3.2 Selected hydrogeologic data on bedrock aquitards

Formation Name, (Location)	Lithology	Test Method	Hydraulic Conductivity K (or T)	Storage, dispersivity, or velocity values	Porosity, fracture apertures, densities	Reference
Pierre Shale, (South Dakota, USA)	Claystone and shale	Lab steady and transient hydraulic, mechanical consolid ©	Steady flow 10^{-12} m/s Transient pulse 10^{-14} to 2×10^{-12} m/s, **C** 10^{-14} to 4×10^{-10} m/s	10^{-6} m^{-1} ; 2×10^{-6} m^{-1} to 10^{-3} m^{-1} (estimated from figure in Neuzil 1986)	0.25-0.4	Neuzil 1986, Neuzil 1994
Pierre Shale, (South Dakota, USA)	Claystone and shale	Inverse anal. of flow systems 0.3x>1km; borehole shut-in, recovery	Transient flow 10^{-14} to 10^{-13} m/s; Boreholes 10^{-15} m/s to 10^{-10} m/s	N/A	0.3-0.33	Neuzil 1993, Neuzil 1994, Neuzil 1995
Eleana Formation (Nevada, USA)	Argillite	Lab hydraulic transient	2×10^{-16} to 10^{-13} m/s	N/A	0.05	Lin 1978 cited in Neuzil 1986, Neuzil 1994
Lower Cretaceous (Western Canada)	Clayey siltstone and clayey sandstone	Lab steady flow	10^{-15} to 2×10^{-12} m/s	N/A	0.15-0.25	Young, Low, and McLatchie 1964 cited in Neuzil 1986, Neuzil 1994
Colorado Gp, Upper Manville Shales (Alberta, Canada)	Claystone, shale	Lg scale inverse anal. of flow systems scale 0.5x>100 km	3×10^{-14} to 5×10^{-13} m/s	N/A	0.2-0.28	Corbet and Bethke 1992 cited in Neuzil 1994
Palfris Formation (Wellenberg Switz.)	Argillaceous marl, interbedded limestone	Field packer, downhole fluid logs dm-m, upscaling simul 100x100x100 m	inflow fractures T = 10^{-12} m^2/s to 3×10^{-5} m^2/s; $K_{eff}<10^{-13}$ to 4×10^{-9} m/s	N/A	0.01-0.03 matrix	Mazurek et al. 1998
Toarcian/ Domerian argillites, (Tournemire, France)	Indurated marls and claystones	Lab (0.1m) and in-situ pulse testing (1m), hydraulic equilibration (10m?)	Pulse decay, triax, HTO 10^{-15} to 10^{-13} m/s; In-situ fractured 10^{-13} to 10^{-11} m/s, unfract. 10^{-15} to 10^{-14} m/s; equilib. 1.4×10^{-14} m/s	Ss=3.1×10^{-6} m^{-1} S=5×10^{-7} to 4.5×10^{-5}	0.02- 0.15	Bonin 1998, Boisson et al. 2001

(continued)

Table 3.2 (Continued)

Formation Name, (Location)	Lithology	Test Method	Hydraulic Conductivity K (or T)	Storage, dispersivity, or velocity values	Porosity, fracture apertures, densities	Reference
Oxford Clay, Kellaways Beds at fault (Down Hampney, UK)	Mudstones	Borehole shut-in pulse and packer tests (1m); simulated porosity	Away from fault $5.1x10^{-12}$ to $5.2x10^{-11}$ m/s; in fault zone $3.1x10^{-11}$ to $2.4x10^{-8}$ m/s	$Ss=4.3x10^{-5}$ to $5.4x10^{-3}$ m^{-1} away from fault; $4.8x10^{-5}$ to $3.2x10^{-4}$ m^{-1} in fault zone	0.05 modeled	Sen and Abbott 1991
Salado Formation (New Mexico, USA)	Evaporites including argillaceous halites, anhydrites, interbedded clay and siltstone	Borehole pressure-pulse, constant-pressure flow and pressure recovery	Total test ranges 10^{-16} to 10^{-9} m/s; Typical averages for Halite 10^{-15} to 10^{-14} m/s; Anhydrite 10^{-13} to 10^{-12} m/s	N/A	0.01	Beauheim and Roberts 2002
Maquoketa Formation (Wisconsin, USA)	Dolomitic shale, interbedded dolomite	Core testing; Borehole pressure, slug and pump testing; inverse model estimates	Lab pulse decay $2x10^{-14}$ to $1x10^{-12}$ m/s; Borehole tests $Kh=3x10^{-10}$ to $3x10^{-5}$ m/s; Fracture T = $8x10^{-6}$ to $6x10^{-5}$ m^2/s; model estimated Kv $1x10^{-14}$ m/s	Lab Ss= $1x10^{-9}$ to $3x10^{-7}$ m^{-1} ; Fracture S = $1x10^{-5}$ to $6x10^{-3}$; simulated S $1x10^{-5}$ to $1x10^{-2}$	Lab 0.03-0.26; simulated fracture 0.001	Eaton 2002

DARCY FLUX AND AVERAGE LINEAR GROUNDWATER VELOCITY

There are two measures of groundwater flow in aquitards: the Darcy flux, often referred to as the seepage velocity, and the average linear groundwater velocity (Grisak et al. 1976). The Darcy flux can be obtained from the well-known Darcy equation (Equation 3.3).

$$q_d = -K[grad\ h] \qquad (3.3)$$

where q_d = Darcy flux (L/T)
K = bulk hydraulic conductivity of the aquitard (L/T)
grad h = gradient of hydraulic head across the aquitard (L/L).

The flux provided by Darcy's Law is a volume of flow per unit cross-sectional area per unit time. The resulting reduced units (L/T) imply velocity, but calculated values using Darcy's Law do not represent the average linear groundwater velocity needed to calculate time of groundwater travel between two locations. Average linear groundwater velocity ($\bar{v}$) is the time required for a water molecule to move downgradient from one point to another divided by the distance between the two points approximated as a straight line. Thus, the average linear groundwater velocity is calculated by Equation 3.4:

$$\bar{v} = K/n\varepsilon\ [grad\ h] \qquad (3.4)$$

where $\bar{v}$ = average linear groundwater velocity
K = bulk hydraulic conductivity of the aquitard
n = porosity effective for groundwater velocity under saturated conditions
ε = an empirical constant
grad h = gradient of hydraulic head across the aquitard.

Experimental data from laboratory and field experiments on sands indicate ε is close to 1 for these materials (Freeze and Cherry 1979). A literature review revealed no information about ε values for water-saturated clayey or silty deposits or for fractured porous media. The usual approach is to assume $\varepsilon = 1$. The effective porosity relevant for calculations of $\bar{v}$ in fractured aquitards is the porosity represented by the interconnected fracture void space, which typically causes equation 3.4 to produce $\bar{v}$ values orders of magnitude larger (e.g., 0.1 to 10 m/day) than the Darcy flux because the fracture porosity values are extremely small. McKay et al. (1993) describe a field tracer experiment in fractured clay where the groundwater velocity is large because of the small fracture porosity.

Grisak et al. (1976) introduced the Cubic Law, adapted from Snow (1969), into the literature concerning fractured clayey aquitards. The Cubic Law provides a basis for quantification of the concept of groundwater velocity in fractured media. However, it assumes a simple conceptual model involving a set of parallel, throughgoing fractures. The intrinsic permeability (k), which has units of L^2, can be calculated for an idealized fractured medium with parallel-plate fractures from Equation 3.5:

$$k = [2/3]\ [1/\Delta]\ \Sigma\ b^3 / N \qquad (3.5)$$

where k = intrinsic permeability
Δ = average fracture spacing
b = half-aperture width
N = total number of fractures in measured area.

Equation 3.5 requires each fracture be uniform in size and orientation If the fracture spacing (Δ) is not known, Equation 3.6, from Snow (1969), can be used to calculate intrinsic permeability:

$$k = [2/3]\ [1/W]\ \Sigma\ b^3 / N \qquad (3.6)$$

where k = intrinsic permeability
W = discharging width
b = half-aperture width
N = total number of fractures in measured area.

The fracture spacing and porosity values given by Grisak and Cherry (1975) were used to calculate an aperture width of 0.0004 cm (40 μm or 0.00016 in) in the clay-loam till at the Whiteshell nuclear research test site in southeastern Manitoba. The fracture spacing in this till is about 4 cm (1.6 in). There are two principal sets of vertical parallel fractures situated roughly at right angles to each other. Using either Equation 3.5 or 3.6, and converting intrinsic permeability to hydraulic conductivity assuming a groundwater temperature of 20°C, a hydraulic conductivity of about $3x10^{-7}$ cm/sec ($1x10^{-8}$ ft/sec) is obtained. This value is close to the conductivity obtained at this site using a combination of pumping test analysis and numerical modeling for steady state groundwater flow (Grisak and Cherry 1975).

From Equation 3.4, smaller fracture porosities cause higher average linear groundwater velocities for a specified gradient. In the case of fractured till, velocities of groundwater in the fractures are many orders of magnitude higher than velocities in the granular matrix. Calculations from Grisak et al. (1976) illustrate this point. A reasonable estimate of the average porosity of fractured till, excluding the intergranular pore spaces in the till blocks between the fractures, is in the order of 10^{-4}, while in the unfractured till the porosity is about 0.30 (Grisak and Cherry 1975). The hydraulic conductivity of the unfractured till determined from consolidation test data is in the order of 5 x 10^{-9} cm/sec ($1.7x10^{-10}$ ft/sec); the bulk hydraulic conductivity of fractured till is one to three orders of magnitude larger, typically in the order of $3x10^{-7}$ cm/sec.

In shallow groundwater flow systems in Pleistocene deposits in the Interior Plains region, hydraulic gradients are commonly in the range of 10^{-1} to 10^{-2}. Using the above ranges of parameter values for fractured till, average linear groundwater velocities calculated using Equation 3.4 are in the range of 0.04 to 4.0 m/year (0.13 to 130 ft/yr). Using the conductivity and porosity data for unfractured till with the same gradient range yields average pore-water velocities in the range of 5 cm (0.16 ft) in 100 to 10,000 years. This indicates water movement through the till blocks between the fractures occurs at an insignificant rate compared to the very active flow system in the fracture networks. Figures 3.14 and 3.15 illustrate these calculations. This large difference between the two types of velocities plays an important role in the migration of contaminants in this type of material. However, the extremely large values for $\bar{v}$ obtained using the calculation procedure outlined above are misleading when considered in the context of the migration velocity of dissolved contaminants in fractured clayey deposits. The calculated $\bar{v}$ is generally much larger than the velocity of a dissolved contaminant front migrating in fractured porous media (silt and clayey aquitards and/or lithified sediments) because dissolved contaminants commonly are strongly influenced by matrix diffusion. Important transport processes for contaminants are discussed in Chapter 4.

OCCURRENCE OF FRACTURES IN AQUITARDS

Historically, hydrogeologists believed fractures in relatively unweathered clayey aquitards were unimportant because of the expectation that natural plasticity would cause fractures to "heal" (e.g., close naturally). Today, open fractures are recognized as abundant in unweathered zones in many aquitards. Early field investigators of groundwater flow in Canada (Lissey 1962, Meyboom 1966, Rozkowski 1967) used piezometers installed in clayey glacial till to show active groundwater flow systems occur in these deposits, but this apparent flow activity was not attributed specifically to fractures. Based on a field study in Illinois, Williams and Farvolden (1967) were the first to identify fractures as the important contributor to hydraulic conductivity and draw attention to the potential for fractures to influence contaminant migration. Their examination of clayey till was prompted by the apparent inconsistency between evidence for active recharge to aquifers and the view that overlying clayey aquitards should be nearly impervious (Norris 1959, Ziezel et al. 1962). In the late 1960s, investigators at the Illinois State Geological Survey incorporated this recognition of fractures in clayey and silty Quaternary aquitards into studies of the subsurface occurrence and fate of contaminants at municipal landfills.

In Manitoba in 1968, a long term investigation of the Whiteshell low-level solid radioactive waste disposal site began in a surficial aquitard composed of clayey glaciolacustrine and clayey till overlying a sandy aquifer. Piezometer measurements identified a very active natural groundwater flow system in the aquitard, previously considered to be nearly impervious (Cherry et al. 1971; Cherry, Grisak, and Clister 1973). The flow-system activity and the relatively large bulk vertical hydraulic conductivity were attributed to fractures. Grisak and Cherry (1975) conducted a long-term pumping test in the aquifer underlying the aquitard while monitoring water levels in many standpipe-type piezometers in both the aquitard and aquifer. The response of piezometers in the fractured aquitard is represented conceptually in Figure 3.13 (Grisak et al. 1976). Although the clay matrix material was plastic and compressible, the fracture network behaved hydraulically as if the aquitard was somewhat indurated (e.g., more like rock than clay). The rapid hydraulic response in the aquitard during the pumping test indicated minimal storativity and compressibility of the fracture network, and showed vertical hydraulic continuity of fractures through the aquitard between the water table and underlying aquifer. This hydraulic response indicated the bulk vertical hydraulic conductivity of the aquitard is in the order of 10^{-7} cm/sec ($4x10^{-8}$ in/sec), about two orders of magnitude greater than values from laboratory tests on core samples. Grisak et al. (1976) summarized the knowledge of the nature and origin of fractures in clayey glacial till in the Plains Region of Canada and the United States up to the mid-1970's, which included information on deep fractures observed in excavations to 20 m (66 ft) below ground surface.

Although the common presence of hydraulically active fractures in clayey Quaternary deposits in the Plains Region and the American Midwest was well established by the late 1970s, information on groundwater velocity and contaminant behavior in these or any other fractured clayey aquitards was lacking.

Early studies in clayey aquitards focused on weathered clayey tills where visual indications of fractures were present. Many surficial clayey aquitards are relatively thick such that deeper in the aquitard weathering features are absent and visual evidence of fractures disappears. In these zones, other types of evidence are needed to determine fracture presence or absence. To determine whether hydraulically-responsive fractures exist in unweathered clay till, Keller, van der Kamp, and Cherry (1986) studied a site near Saskatoon, Saskatchewan, referred to as the Dalmeny site. At this site, a sand aquifer is directly overlain by 7 m (23 ft) of unweathered gray clay till, in turn overlain by 12 m (39 ft) of brown weathered clayey till to ground surface. Vertical borehole samples from the unweathered gray till showed no visual evidence of fractures. The aquifer was pumped for 3 days, during which time the hydraulic response of the gray and brown till zones were monitored using a piezometer network. The decrease in hydraulic heads moved quickly upward through the unweathered till into the weathered till, demonstrating that although the unweathered till exists at considerable depth below the water table, the till is nevertheless hydraulically responsive. Because the response time was faster than would be possible if the unweathered clayey zone was behaving according to matrix properties (i.e., lab determined parameters), the only plausible cause of the responsiveness is open vertical fractures. This example indicates absence of visual features of weathering in a clayey aquitard sampled by vertical boreholes cannot be taken as sufficient evidence to conclude there is a lack of hydraulically-active fractures. For a corehole to show visual evidence of a fracture, it must encounter the fracture directly and the fracture much have color features due to weathering. Fractures may exist without such coloration and therefore not be identifiable. The advantage provided by pumping tests with monitoring within the aquitard is the piezometer does not have to be connected directly to the fracture to provide a response indicative of a fracture; the pore-pressure response will propogate away from each fracture to encompass ever-increasing zones of pore-pressure decline while the pumping test is in progress.

Although bedrock aquitards exert strong influence on regional flow systems, the study of bedrock aquitards is much less advanced than the study of surficial clay aquitards. Further complicating the problem, where bedrock aquitards such as shale are exposed at land surface or in outcrops, they are often highly weathered and do not have the same appearance as the unweathered rocks present in the subsurface. An example of a highly weathered shale is the semi-lithified material called sapolite (McKay, Sanford, and Strong 2000). Weathering has caused this material to no longer exhibit aquitard properties. Some bedrock aquitards, such as the Opalinus Clay in Switzerland (Gautschi 2001) and Toarcian argillites in France (Patriarche et al. 2004a, Patriache et al. 2004b), although known to be potentially fractured and faulted, nevertheless have such low effective hydraulic conductivity they are being assessed for repositories for radioactive waste.

Bedrock aquitards are often composed of beds of differing minerology such as shale, dolomite or chalk (Eaton 2002, Rijken and Cooke 2001), or beds showing variations in lithology such as shale, mudstone and sandstone (Bishop et al. 1993a). Formations composed mostly of mudstone with intervals of sandstone have sometimes been used as multiple-aquifer systems where deep wells with numerous perforated intervals pump primarily joints from the sandstone beds (Lerner, Burston, and Bishop 1993). In layered sedimentary aquitards, fractures form perpendicular to the more brittle beds, such as sandstone, chalk or dolomite, and fractures also occur parallel to planes of weakness between different lithologies (e.g., bedding partings). The vertical or near vertical fractures often terminate on more ductile beds such as shale (Rijken and Cooke 2001) or mudstone (Lerner, Burston, and Bishop 1993) that are less fractured. Bedding-plane fractures in some lithified aquitards are enlarged by mineral dissolution, and may form highly transmissive zones (Barrash and Ralston 1991, Eaton 2002, Michalski and Britton 1997).

Fracture networks in layered sedimentary rocks are often highly anisotropic with limited vertical interconnection between some beds or bedding planes. Groundwater flow patterns in such layered bedrock aquitards can be very different than reported for non-indurated aquitards where vertical flow is the normal expectation. Lateral flow in such fractured bedrock aquitards can control the hydraulic head distribution, particularly near lateral boundaries (Eaton 2002). Monitoring of the vertical head distribution in combination with monitoring in an underlying pumped aquifer system has been used to estimate effective vertical hydraulic conductivity of bedrock aquitards (Eaton and Bradbury 2003). Because of such highly anisotropic fracture networks, silt- and sandstone formations considered aquitards in the context of vertical flow between major aquifers can be sources for springflow due to horizontal flow allowed by the much larger horizontal component of hydraulic conductivity (Swanson 2002).

CROSS-CONNECTING BOREHOLES IN AQUITARDS

In many parts of the United States, water supply wells have screened sections or open borehole sections connected to more than one aquifer. In general, these multi-aquifer well completions are efficient for achieving larger well yields, but often have the unintended consequence of providing hydraulic connection between aquifers naturally separated by aquitards. Thus water from one aquifer will "short circuit" into an adjacent aquifer if there is a head differential between the aquifers causing contaminants or other solutes from one aquifer to enter the other. Therefore, multi-aquifer wells pose a threat to aquitard integrity (Lacombe et al. 1995, SEWRPC/WGNHS 2002). In a complex deep mudstone-sandstone system underlying the City of Coventry, UK, cross-connecting boreholes caused significant increases in chlorinated solvent contamination measured in underlying aquifer units (Bishop et al. 1993b). In this type of setting where numerous water supply wells have intake zones across multiple aquitard and aquifer units, the sources of groundwater contamination and its subsequent transport pathways are difficult to identify. Even in wells reported to be cased across aquitards there is a potential for rapid contaminant migration along the well annulus outside of the casing if the well grout is missing, poorly emplaced, deteriorated, or fractured. Unfortunately there are few reliable methods for evaluating grout seals outside of steel well casing.

Figure 3.16 illustrates the situation of wells connecting two aquifers with different hydraulic heads. Coupling the flow equations for the two aquifers allows calculation of the rate of flow down the well bore from the upper to the lower aquifer. Using the values typically representative of the flow system in southeastern Wisconsin, Hart, Bradbury, and Feinstein (N.d.) estimated the flow through a single multi-aquifer well to be 1000 m^3/day (2.6 x 10^{-5} gal/day). Relatively few such multi-aquifer wells are required to increase considerably the large-scale K_v of an aquitard.

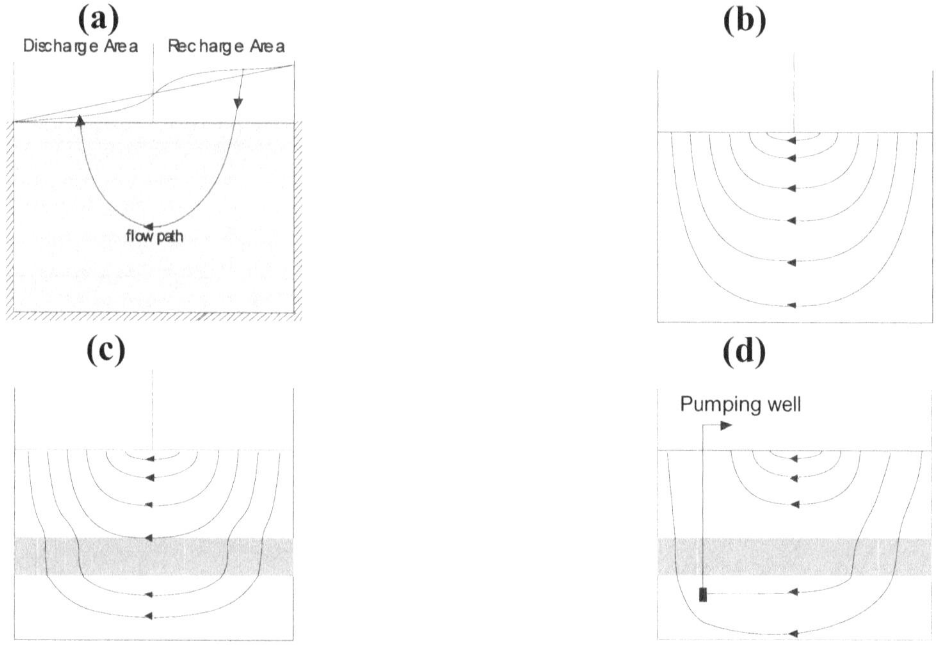

Modified from Zijl and Nawalany 1993.

Figure 3.1 Influence of aquitard on groundwater flow lines in a simple bounded steady state flow system (a) flow domain and boundary conditions (b) continuous aquifer (c) single continuous aquitard between two aquifers (d) pumping well in lower aquifer

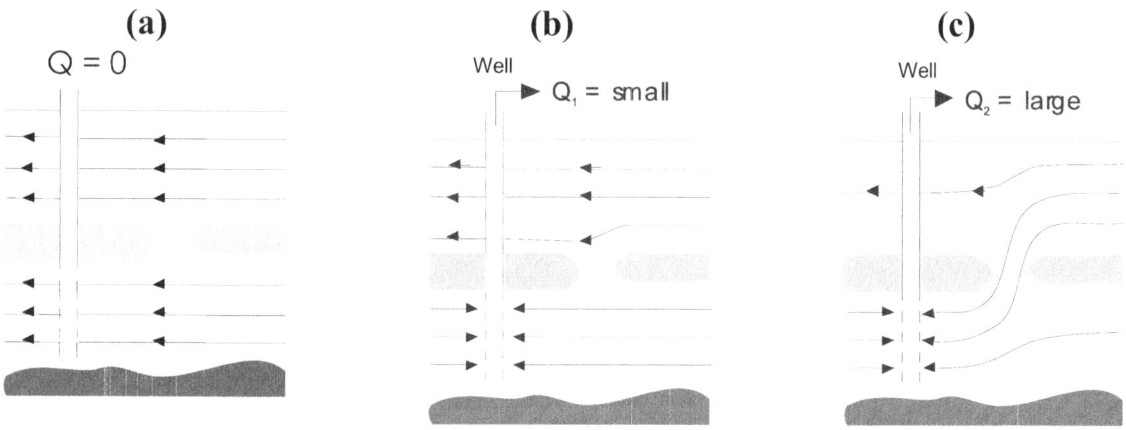

Modified from Zijl and Nawalany 1993.

Figure 3.2 Groundwater flow paths in the vicinity of a window: (a) no pumping (b) weak pumping drawing minimal water from the upper aquifer (c) strong pumping inducing strong flow from the upper aquifer

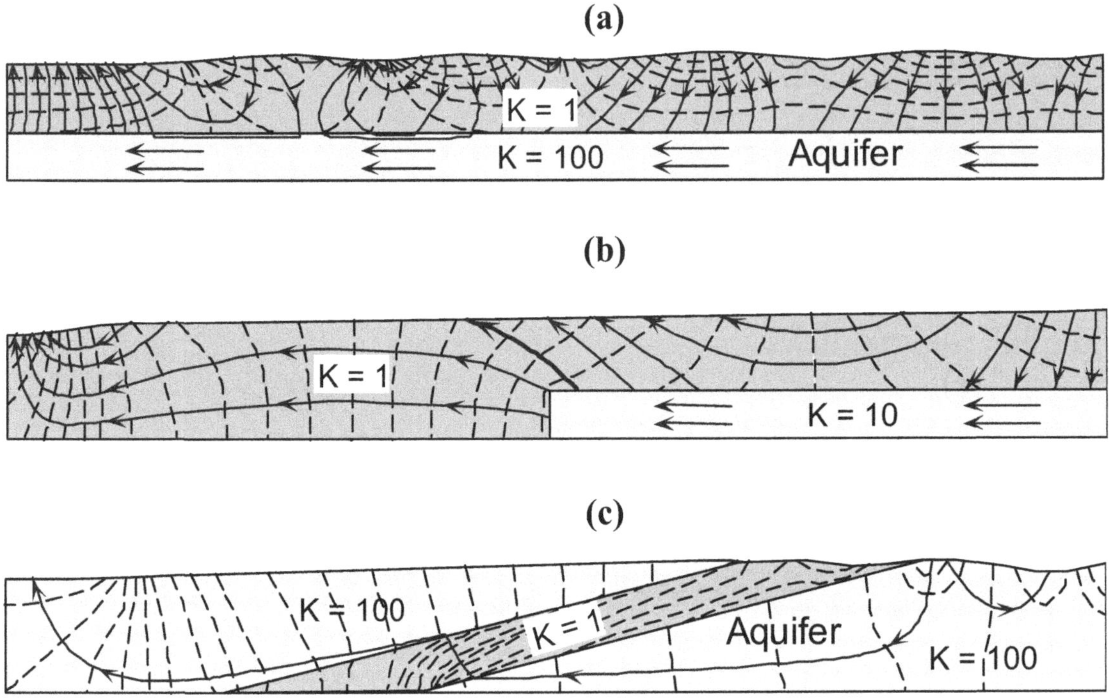

Source: Freeze and Witherspoon 1967

Figure 3.3 Effect of geology on regional groundwater flow path produced using a two dimensional numerical model

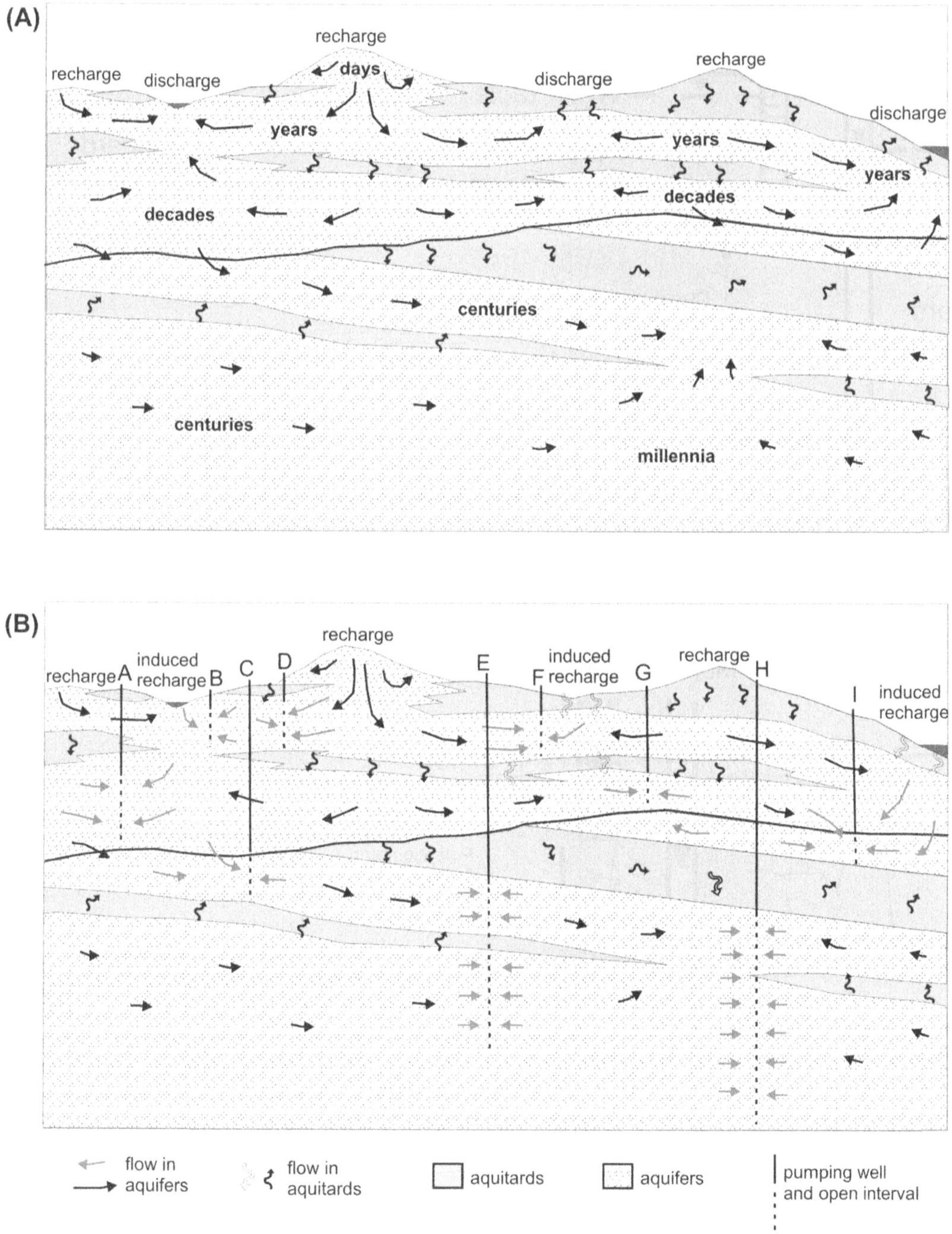

Figure 3.4 Effects of pumping wells on regional groundwater movement (a) regional groundwater flow system in overburden and bedrock under natural conditions (b) the flow system is disturbed due to pumping wells.

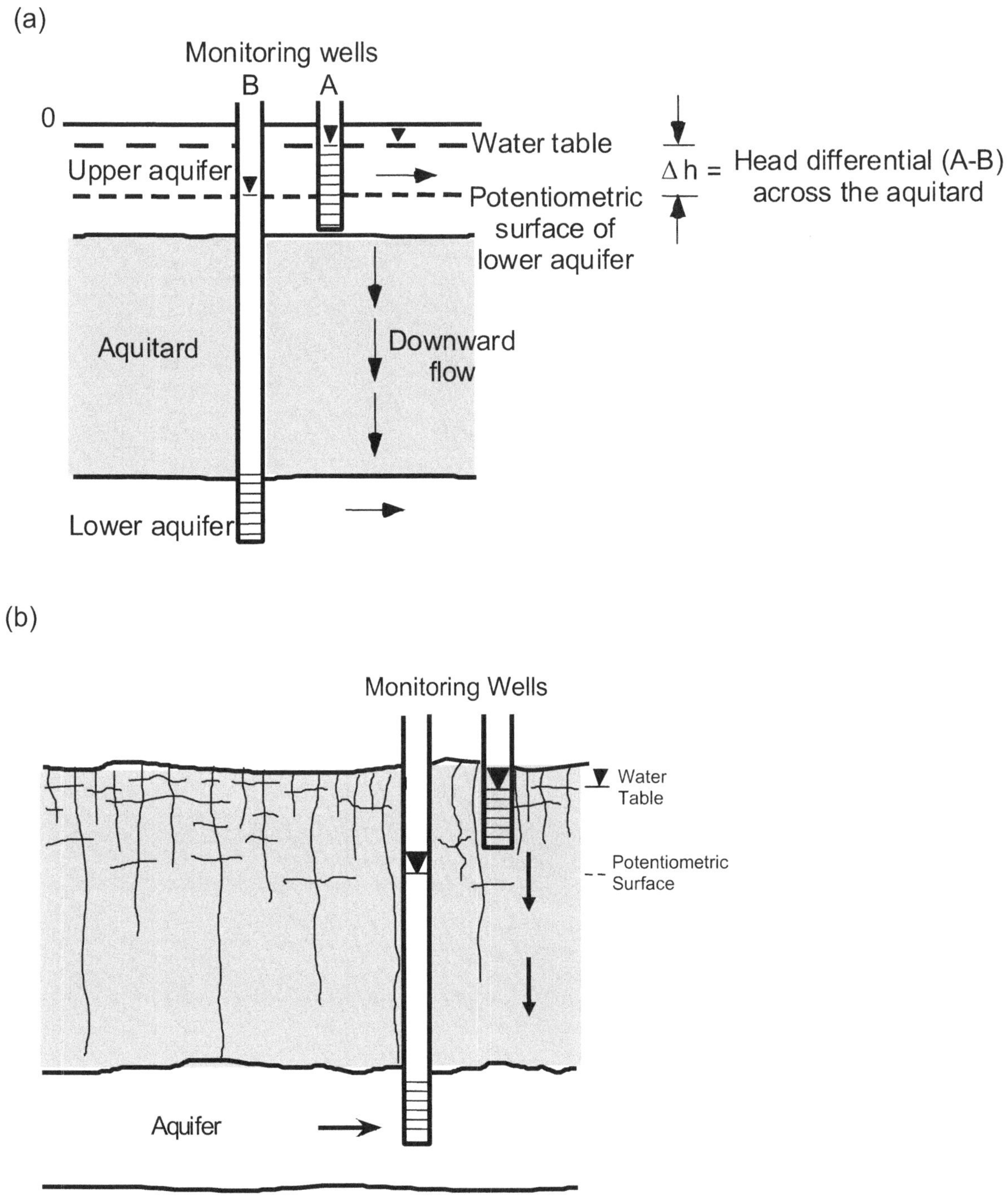

Figure 3.5 Determination of the hydraulic gradient across aquitards using head measurements external to the aquitard: (a) aquitard between two aquifers where head is measured in the two aquifers (b) surficial aquitard with shallow water table in fractured zone overlying aquifer.

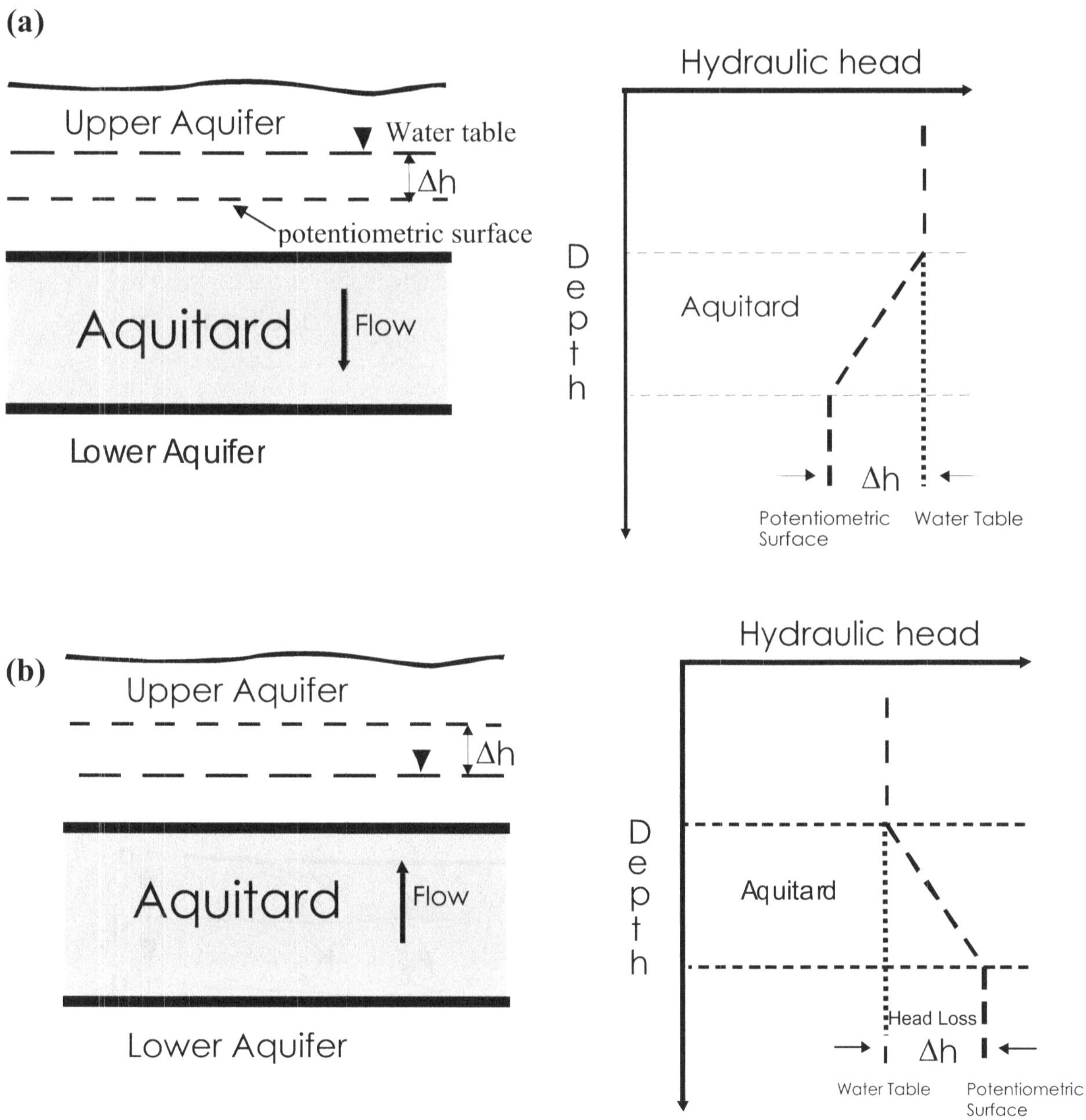

Figure 3.6 Schematic example of two cases of vertical flow across homogeneous aquitards: (a) downward flow and (b) upward flow.

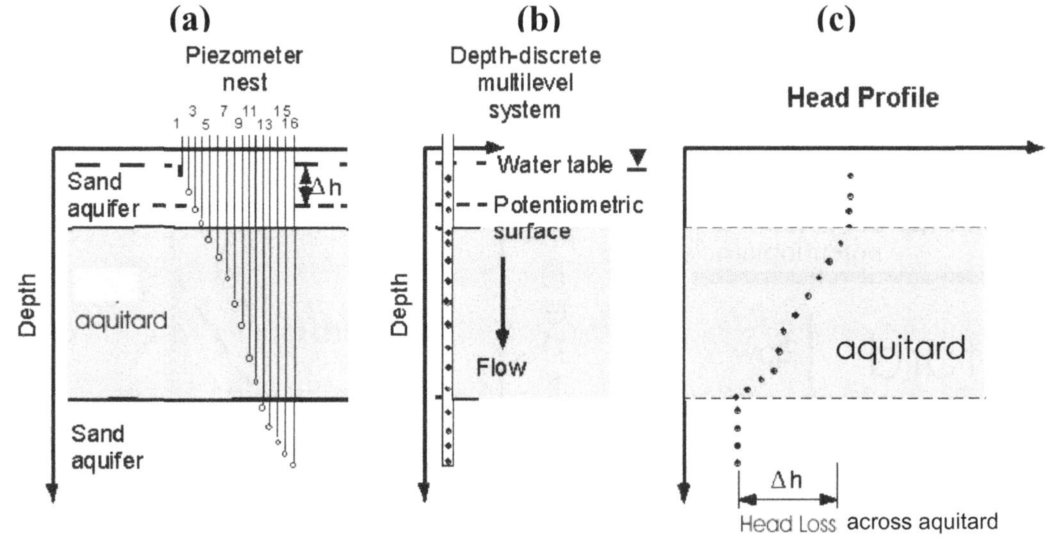

Figure 3.7 Schematic example of a detailed profile of hydraulic head verses depth in an aquitard obtained by multilevel monitoring: (a) piezometer nest (b) multilevel monitoring system (c) head profile.

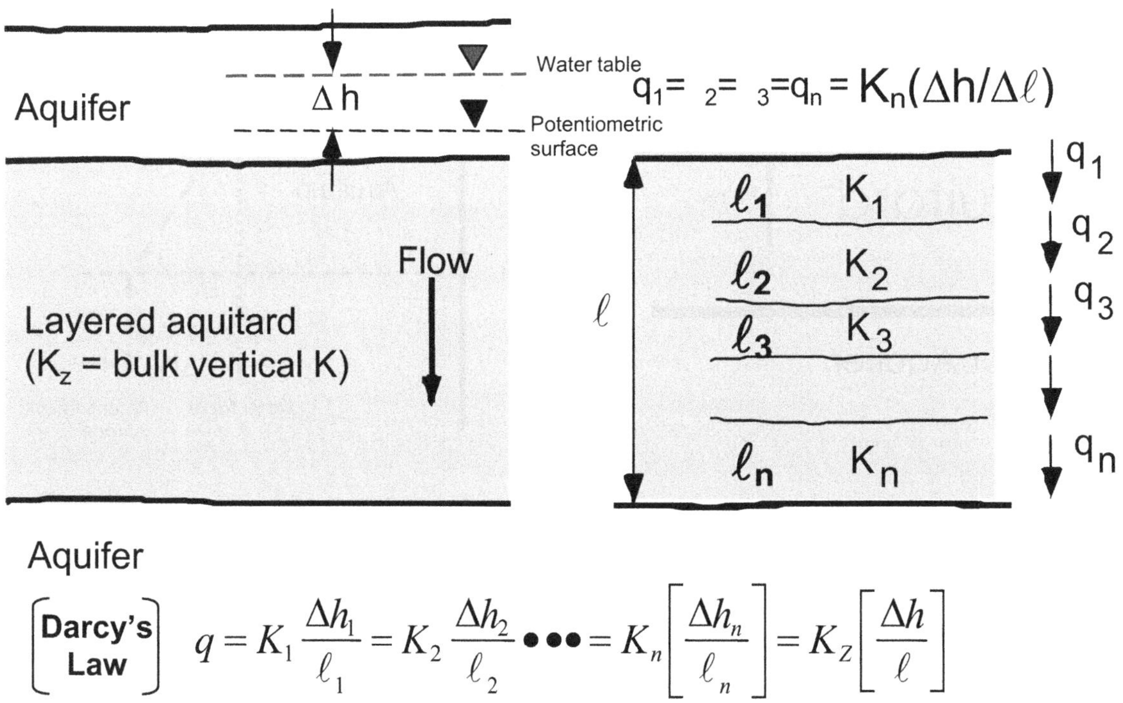

Figure 3.8 Application of Darcy's Law to represent continuity of vertical groundwater flow in a layered aquitard. Assuming that there is only vertical flow in the aquitard, the Darcy flux across all layers is the same

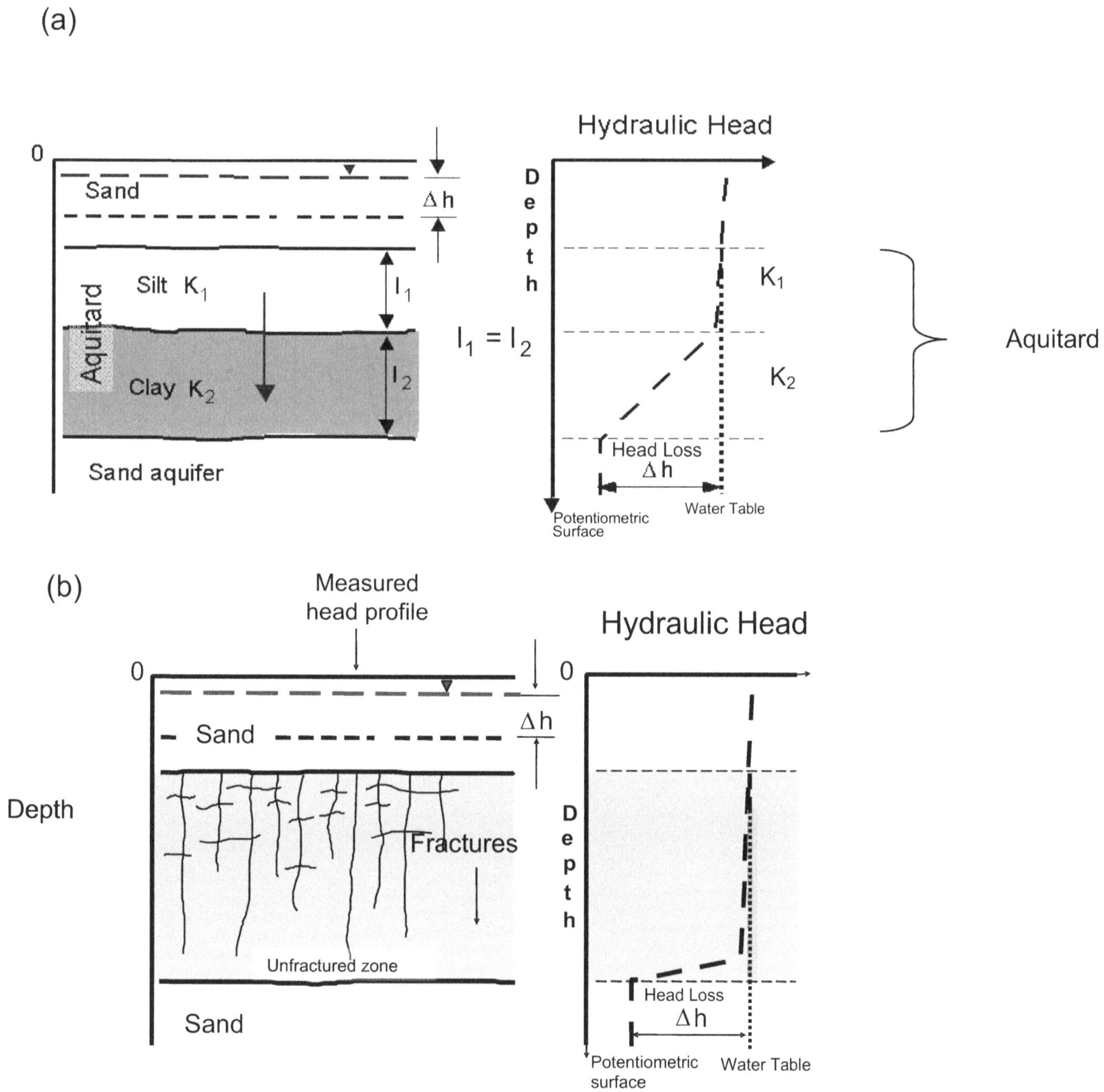

Figure 3.9 Hydraulic head distribution within an aquitard with two zones (K1 >> K2): (a) upper zone is silt and lower zone is clay where nearly all of the head loss occurs (b) upper zone is fractured clay and the lower zone is unfractured clay where nearly all of the head loss occurs

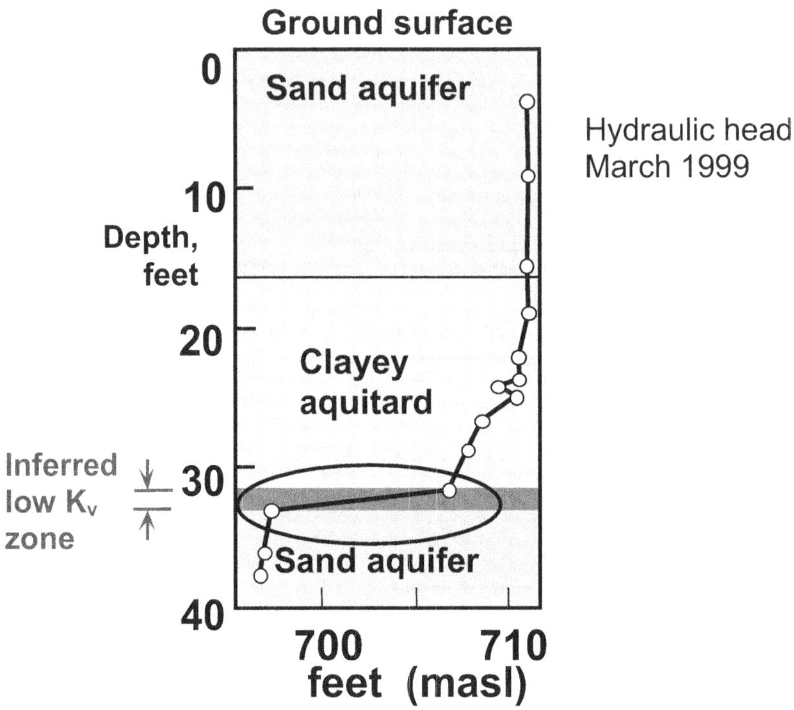

Source: Meldrum 1999

Figure 3.10 Vertical hydraulic head profile measured in a clayey aquitard using a multilevel monitoring system. Nearly all head loss occurs at the bottom of the aquitard indicating the vertical hydraulic conductivity is lowest in this zone. Core logs indicate that the variability in the head profile cannot be attributed to textural variations and therefore fractures are the inferred cause.

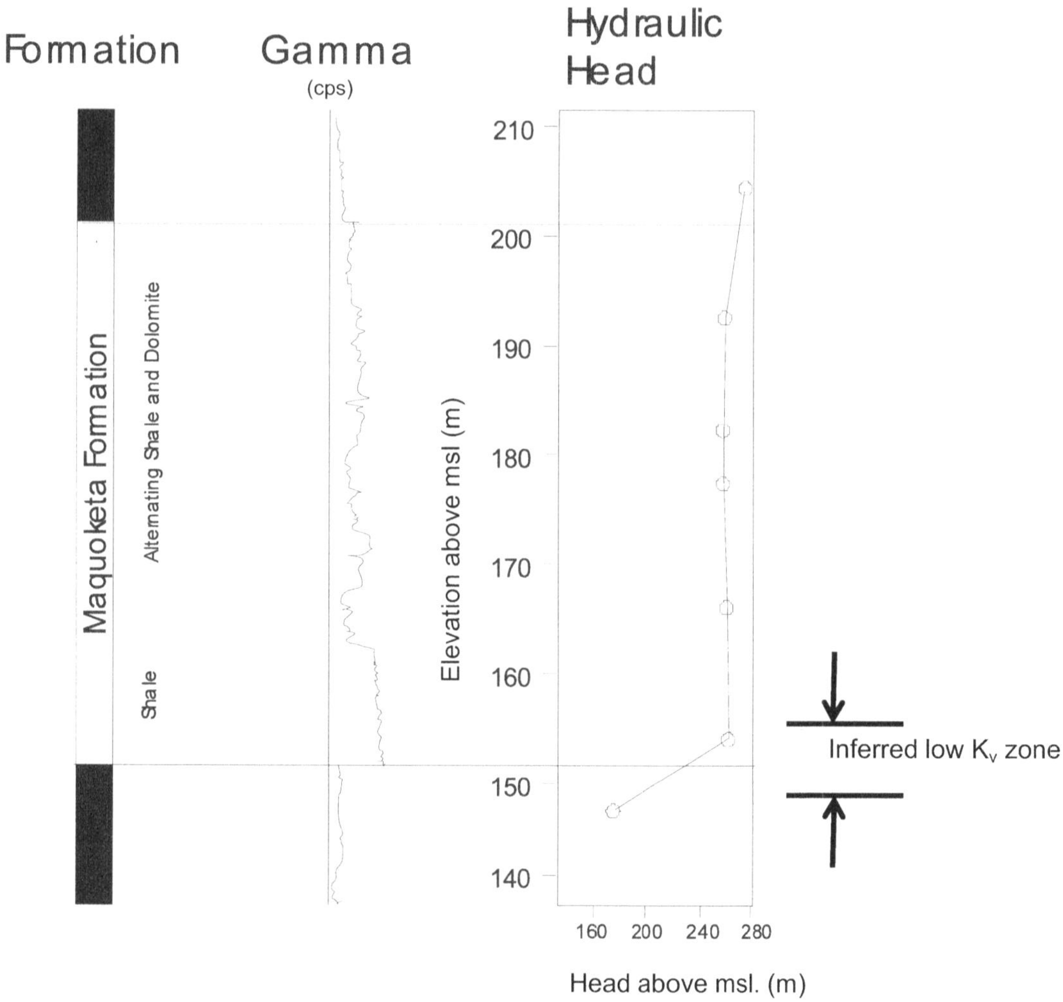

Source: Adapted from Eaton, 2002

Figure 3.11 Vertical hydraulic head profile across a shale aquitard in Wisconsin (Maquoketa Formation). Nearly all head loss occurs at the bottom of the shale indicating the vertical hydraulic conductivity is lowest in this zone.

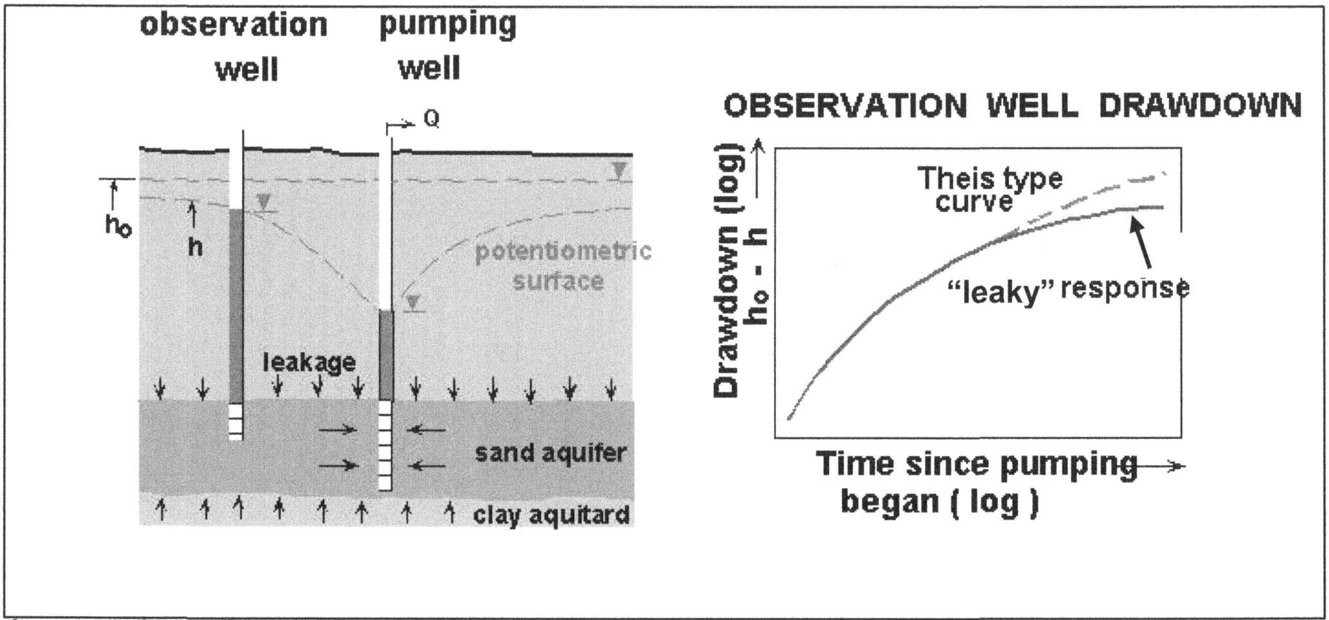

Source: Adapted from Parker and Cherry, University of Waterloo, Earth 653 Course Notes

Figure 3.12 Conventional pumping test monitor and leaky aquitard response

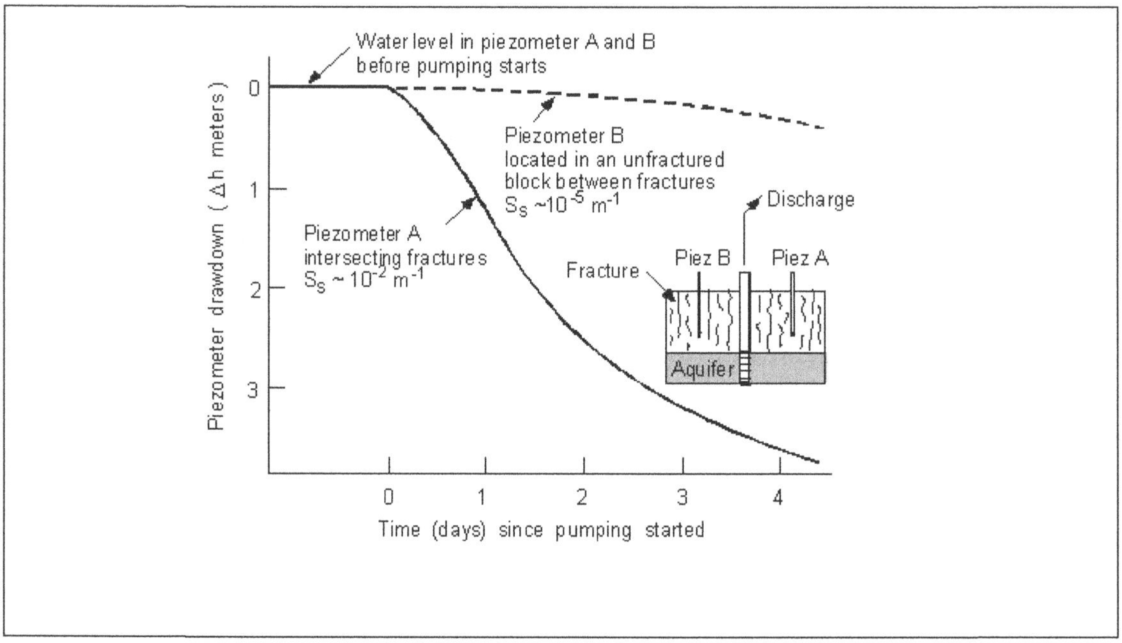

Source: Grisak et al. 1976

Figure 3.13 Greatly different drawdown responses from two piezometers in a fractured aquitard during an aquifer pumping test: one piezometer (A) connected to a hydraulically active fracture shows rapid drawdown while another piezometer (B) not connected to a fracture shows almost no drawdown

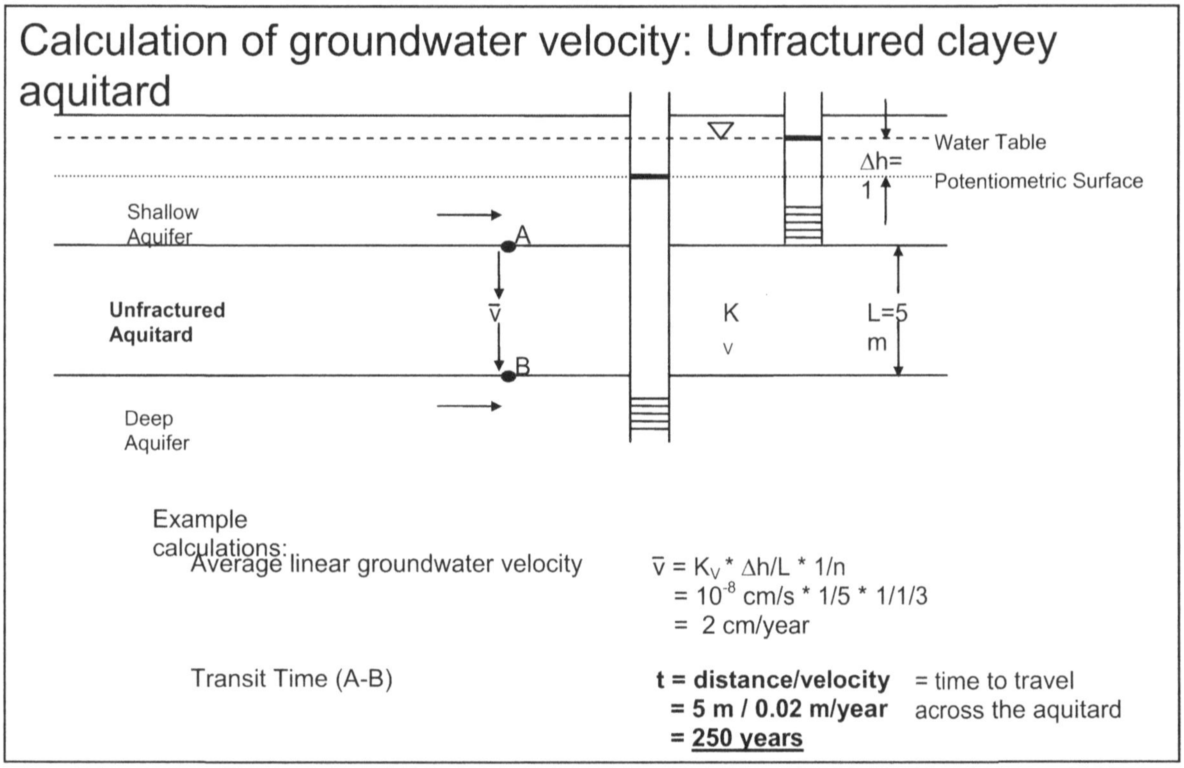

Figure 3.14 Example calculation of groundwater velocity and travel time across an unfractured clay aquitard (5 m thick). Calculation based on Darcy's Law

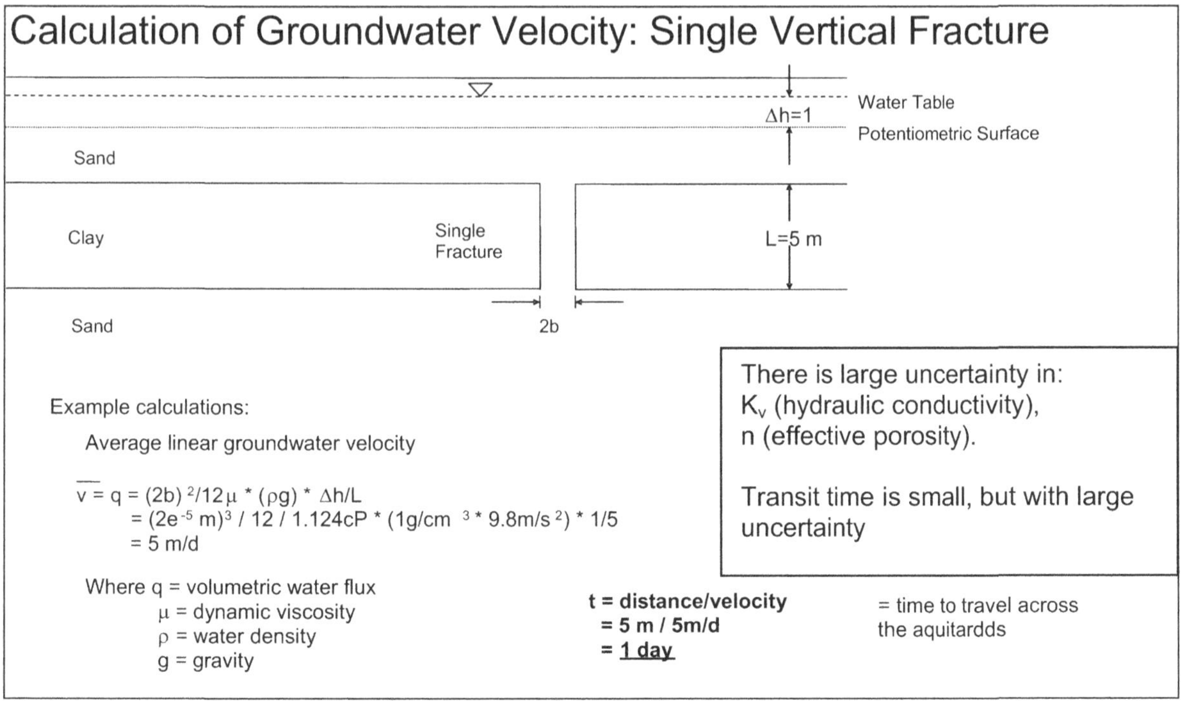

Figure 3.15 Example calculation of groundwater velocity and travel time across a fractured aquitard (5 m thick). Calculation based on Cubic Law for medium with parallel smooth vertical fractures.

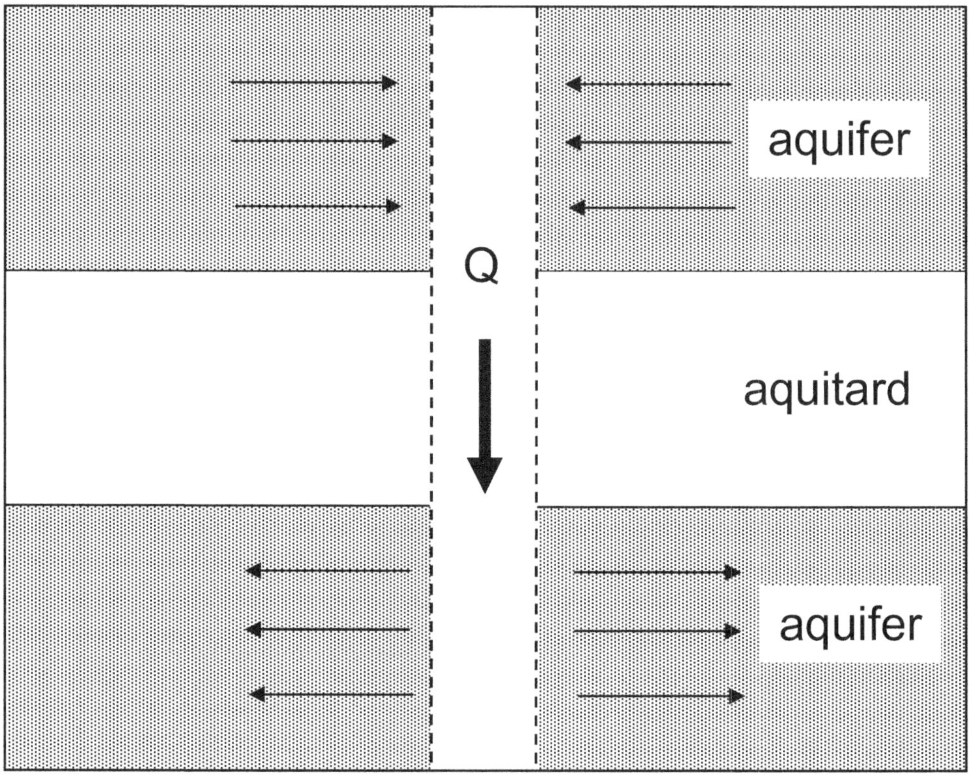

Figure 3.16 Flow from an upper aquifer to a lower aquifer via an open cross connecting borehole where Q is flow

CHAPTER 4

CONTAMINANT MIGRATION THROUGH AQUITARDS

FRAMEWORK FOR CONSIDERATION OF CONTAMINANT MIGRATION

Aquitard integrity depends on the capability of the aquitard to prevent, delay or strongly attenuate the flux of contaminants into an underlying aquifer. Assessment of aquitard integrity involves development of predictions of contaminant migration through the aquitard. For the predictions to be most useful for groundwater resources management, they should include the contaminant arrival time and the subsequent breakthrough curve (e.g., concentration versus time) for the contaminant entering the aquifer and also the contaminant discharge versus time (i.e., the loading curve) for the contaminant mass entering the aquifer. Figure 4.1 illustrates these concepts. The concentration breakthrough and loading graphs in this figure pertain to a specified point (X) and a horizontal area (LW) at the interface between the aquitard and the lower aquifer where the plume enters the lower aquifer. The impact of the plume on any pumping wells in the lower aquifer can be represented by similar shaped graphs. Although the objective is to quantitatively predict the future concentration and loading, uncertainties in various components of the analysis are typically so large the endeavour is at best semi-quantitative and often involves "orders-of-magnitude" estimates.

Field data needed to develop reliable predictions are commonly unattainable because of funding limitations and/or unquantifiable complexities in the hydrogeologic system. Therefore, decisions are often made based on an information framework with combinations of quantitative and qualitative data with major gaps. Judgments are required to prioritize different hypotheses or conceptual models for which supporting evidence is available. In one conceptual model, contaminants may arrive at a well within a predictable time period, yet in the competing hypothesis they will never arrive. The main uncertainty can pertain to the selection of the most representative conceptual hydrogeologic model rather than the details of model parameter values.

Inherent in investigations of aquitard integrity is the need to acquire field data indicating whether or not important preferential pathways exist in the aquitard. The primary challenge in predicting contaminant migration is to determine the nature of contaminant migration pathways, such as intergranular flow versus flow through fractures, because this determination dominates the predicted breakthrough and loading curves. For example, the predicted time for passage of contaminants through a thick unfractured clayey aquitard is typically hundreds or thousands of years, but for an aquitard with substantial occurrences of fully penetrating open fractures, the passage time could be orders of magnitude smaller. Water-management decisions are normally not influenced by the magnitude of the uncertainties if the timescales all involve future centuries, but if the uncertainties pertain to what will happen next year or next decade, they can have great importance.

If the aquitard or a part of the aquitard under investigation likely has no preferential pathways (i.e., no fractures, no root holes and other potential fast-flow openings), the predictive endeavor can then focus on intragranular contaminant migration through the low-permeability porous media. Section 4.2 describes migration of dissolved contaminants in low-permeability aquitards with no preferential pathways, where molecular diffusion commonly dominates contaminant migration. If the aquitard has or could have preferential pathways such as fully penetrating fractures, then models should incorporate advection in the fractures with influence of diffusive transfer of contaminant mass between the fractures and the low-permeability matrix material. Sections 4.3 and 4.4 consider this type of contaminant migration in aquitards.

The best prospects for developing reliable predictions of migration of dissolved contaminants in aquitards in future decades or longer occur when actual contaminants or contaminant analogs such as environmental isotopes (e.g., atmosphere-derived tritium, oxygen-18, deuterium) already exist in the aquitard. Sections 4.5 and 4.6 provide examples in which this approach provides much of the basis for determination of aquitard integrity.

DIFFUSION DOMINATED UNFRACTURED AQUITARDS

Most studies of contaminant migration have focused on aquifers where advection is the primary driving mechanism. However in aquitards, molecular diffusion, hereafter referred to simply as diffusion, is an important process influencing contaminant migration and in some cases is the dominant process. This is typically the case in unfractured clayey aquitards where the groundwater velocity can be so small Fick's First Law for diffusion rather than Darcy's Law for flow provides the basis for consideration of the contaminant migration rate. Narasimhan (2004) describes the origin of Fick's Law in chemistry in 1855, a year before Darcy published the experimental results establishing Darcy's Law for groundwater flow. Foster (1975) was the first to draw attention to the importance of Fick's Law in contaminant hydrogeology, which pertained to retardation of solute migration in fractures relative to groundwater flow rate.

Using a mathematical analysis, Gillham and Cherry (1982) assessed the relative influences of diffusion and advection on the migration of non-reactive contaminants in unfractured aquitards. The advection-dispersion equation for one-dimensional transport of dissolved constituents in a homogeneous saturated porous medium is:

$$\frac{\partial C}{\partial t} = D_\ell \frac{\partial^2 C}{\partial \ell} - \bar{v}\frac{\partial C}{\partial \ell} \qquad (4.1)$$

where: C represents the solute concentration

t is time

l is distance along the migration path

D_l is the dispersion coefficient in the migration direction

$\bar{v}$ is the average linear groundwater velocity

An analytical solution to equation 4.1 for an infinite travel path and a step-function solute input was used for calculating profiles of solute concentration versus distance for specified values of t. In these calculations the range in velocity was always small enough for the dispersion process to be dominated by molecular diffusion and thus D_l was set equal to the effective diffusion coefficient (D_d). Velocities sufficiently small for diffusion to be dominate over dispersion are typical of unfractured clayey deposits. The set of results in Figure 4.2 indicates low groundwater velocities have little influence on the advance of the solute front. For example, in Figure 4.2a, at low velocities, a tenfold increase in velocity (0.32 to 3.2 cm/yr) results in a 50% increase in the front advance at $C/C_o = 0.5$. In the preparation of these graphs, diffusion coefficients of 1×10^{-5} and 1×10^{-6} cm^2/s were used. These values span the normal range reported in the literature for nonreactive solutes in saturated, unconsolidated deposits such as silt or clay.

Figure 4.3 provides a further comparison of the effect of diffusion relative to advection, and shows the rate of advance of the solute front, calculated as for Figure 4.2, versus the rate of advance that would occur by advection alone (that is, plug displacement of the solute front). In the development of Figure 4.3, the rate of front advance was calculated using the $C/C_o = 0.5$ and 0.01. Devlin and Parker (1996) describe the theoretical circumstances in low-permeability under which dissolved chemicals can migrate media by diffusion against the hydraulic gradient.

Graphs such as Figures 4.2 and 4.3 are useful in a general way to identify when diffusion is important. For example, assume 10^{-2} as a representative hydraulic gradient for a particular field setting, 0.35 as a representative porosity, and 1×10^{-5} cm^2/s as a representative diffusion coefficient for a nonreactive contaminant. Given these parameters, Figure 4.3 indicates contaminant migration is controlled by diffusion in deposits with hydraulic conductivity less than about 10^{-6} cm/s; when the conductivity is greater than about 10^{-5} cm/s, it is controlled by advection. In the range between these conductivity values, both advection and diffusion are important.

The most reliable investigative approach for determining whether contaminant migration in a particular aquitard is diffusion-controlled is the analysis of solute profiles. Such analyses are possible if the aquitard already has contamination. Comparisons between the vertical contaminant distribution in the aquitard and simulations of contaminant migration from analytical or numerical models can help determine the importance of diffusion. Goodall and Quigley (1977) were the first to obtain contaminant concentration profiles from a clayey aquitard and consider diffusion-based transport in the data interpretation. Crooks and Quigley (1984), Liu and Ball (2002), Johnson, Cherry, and Pankow (1989) and Parker, Cherry, and Chapman (2004) also provide examples of this approach for clayey aquitards. Crooks and Quigley (1984) studied diffusion profiles for NaCl beneath a municipal landfill in the thick Sarnia, Onatio area and projected 3.5 m of downward migration over 100 years. Johnson, Cherry, and Pankow (1989) collected vertical cores from an aquitard beneath a deep disposal pit at an industrial waste landfill also in the Sarnia area clay. These profiles showed chloride and trichloroethylene concentrations also closely matched mathematical simulations of diffusive contaminant migration with no advection. In Connecticut, Parker, Cherry, and Chapman (2004) obtained trichloroethylene profiles from below DNAPL lying on top of a clay-silt aquitard. These profiles closely matched diffusion dominated mathematical simulations with a small influence of advection. They predicted hundreds of years will pass before the contaminants will reach the aquifer beneath the aquitard. The aquitards at these sites are thick (>20 m unfractured), glaciolacustrine deposits and the simulations predict many hundreds of years or longer will pass before the diffusion-dominated contaminant migration causes breakthrough into the underlying aquifers.

Many aquitards for which integrity assessments are needed do not yet have contaminant migration, and therefore, other approaches are needed to produce contaminant migration estimates. Thick clayey aquitards in which the groundwater flux is relatively small commonly exhibit distributions of natural dissolved inorganic constituents or isotopes of water (e.g., ^3H, ^{18}O, ^2H) suitable for estimating the migration rates of all non-reactive dissolved constituents including many types of dissolved contaminants. The migration rates for non-reactive constituents are faster than those of all other constituents and therefore provide upper limits.

INFLUENCE OF MATRIX DIFFUSION ON CONTAMINANT MIGRATION IN FRACTURED AQUITARDS

As indicated in Chapter 3, field studies provide evidence of the occurrence of natural fractures in many aquitards. Acquisition of evidence of fractures is an important initial step in assessment of aquitard integrity. However, fracture occurrence alone is not a basis for concluding an aquitard has poor integrity. Several hydrogeologic factors govern contaminant migration in fractures, including continuity of fractures (e.g., fully or partially penetrating), fracture apertures, hydraulic gradient and matrix diffusion, and the type of contaminant. Dissolved contaminants are an appropriate starting point for the consideration of contaminant migration and fate in fractured aquitards, with initial consideration focused on transport by the bulk motion of the flowing groundwater.

The large values for calculated average linear groundwater velocity in fractures presented in chapter 3 give the erroneous impression contaminants dissolved in the groundwater (i.e., solutes) will always travel quickly through the fracture network. However, based on studies of the Chalk aquifer in England, Foster (1975) showed the advancement of a contaminant front moving along fractures strongly retarded relative to the average linear groundwater velocity in the fractures. Figure 4.4 illustrates this conceptual model derived from Foster (1975) for contaminant retardation in idealized parallel-plate fractures. The conceptual model is exemplified by a simple mathematical model involving a single planar fracture in a low-permeability porous medium with the diffusion-driven mass transfer of a non-reactive solute from the advecting flow regime in the fracture into the matrix where the pore-water is immobile. This loss of contaminant mass from the fracture causes the contaminant front to move slower relative to the plug-flow position of a hypothetical front advancing at the rate of the average linear groundwater velocity. Building on the work of Foster (1975), Day (1977) applied this retardation concept to solute migration in a fractured clayey aquitard. Day developed and applied an analytical-numerical model for solute behavior in a set of parallel fractures to explain the distribution of atmosphere-derived ^{3}H and Pleistocene-age ^{18}O and ^{2}H in a glaciolacustrine aquitard in Winnipeg, Manitoba. The mathematical models developed by Foster (1975) and Day (1977) have limited capability, which prompted development of more robust models, such as the analytical models of Tang, Frind, and Sudicky (1981) and Sudicky and Frind (1982), the numerical model of Grisak and Pickens (1980) and most recently the numerical model of Sudicky and McLaren (1992). Grisak, Pickens, and Cherry (1980) describe the application of their numerical model in the interpretation of column experiments involving solute transport through a large, relatively undisturbed cylindrical sample of fractured clayey till from the Whiteshell nuclear research site. This till had been shown (Grisak and Cherry 1975) to have fully penetrating hydraulically active fractures.

Independent of the investigations conducted in clayey aquitards by hydrogeologists in the Plains region of Canada, geotechnical engineers, most prominently R.M. Quigley and colleagues at the University of Western Ontario, were investigating the geotechnical properties of contaminants and water isotope migration from municipal landfills in the thick surficial clayey aquitard in the Sarnia area of southwestern Ontario. The surficial aquitard underlying the City of Sarnia extends eastward from the St Clair River that forms the Ontario-Michigan border to more than 100 kms into Ontario and also westward from the river into Michigan. This clayeyaquitard, originating in late Pleistocene pro-glacial lakes, is 30-50 m thick near the St Clair River and overlies a thin regional aquifer. The City of Sarnia has a large petrochemical industry and several landfills. The integrity of this aquitard has been the subject of many field studies since the initial work by Goodall and Quigley (1977), who used vertical profiles of inorganic contaminants in the natural clay beneath landfills to conclude that the vertical fractures caused contaminants from the bottom of landfills to have migrated deeper in the clay than would otherwise have been the case. This was the first of many studies by Quigley and colleagues of diffusion-influenced contaminant migration and fate in the natural clay beneath landfills in this area (e.g., Barone, Rowe, and Quigley 1990; Barone et al. 1989; Fernandez and Quigley 1985; Fernandez and Quigley 1988; Fernandez and Quigley 1991; King et al. 1993; Quigley, Fernandez, and Rowe 1988; Quigley et al. 1987). Johnson, Pandow, and Cherry (1989) also reported on contaminant migration in the Sarnia area aquitard.

Starr, Gillham, and Sudicky (1985) conducted a lab experiment in which a thin sand layer was sandwiched between two silt beds in a box. The sand layer represented a single permeable layer with rapid groundwater flow. When solute tracers were added to water flowing in the sand bed, matrix diffusion from the sand layer into the silt caused strong retardation. This experiment was used to validate an analytic mathematical model, with the sand layer being analogous to a single parallel-plate fracture and the silt representing the matrix material.

Sudicky and McLaren (1992) developed a two-dimensional mathematical model (FRACTRAN) to represent steady-state groundwater flow and transient solute transport in discretely fractured porous media. The capability of this model for representing solute transport in fractured media greatly exceeds those of previous models. FRACTRAN allows the fracture network to have a large number of orthogonal fractures with variable fracture lengths and apertures. Thus, many of the features of actual fracture networks in the field can be represented in the model. The model simulates advection and diffusion in the fractures, and diffusion and advection in the matrix blocks between fractures. FRACTRAN has proven useful for representing idealized fractured systems for purposes of examining the importance of various parameters or network features. Harrison, Sudicky, and Cherry (1992) used FRACTRAN to examine the influence of several factors on the downward migration of dissolved phase trichloroethylene (TCE) through an idealized 15 m (49 ft) thick water-saturated clayey aquitard overlying a horizontal sand aquifer. The fracture network was comprised primarily of vertical fractures with increased lateral spacing at greater depth representative of contractional-type (e.g., dessication) fractures typical of surficial aquitards. The fracture network had many fractures but only a few vertical fractures penetrated the full distance from the contaminant input zone to the bottom of the aquitard. In the model, TCE was released as a dissolved contaminant near ground surface and was transported downward through the fracture network. The model simulation showed the few fully penetrating fractures with apertures > 10 μm caused TCE to pass through the aquitard, resulting in contamination of the underlying aquifer within a few decades. The downward leakage of contaminants through the few deep fractures in the aquitard produces a thin horizontal plume in the upper part of the aquifer. For context, the typical diameter of human hair is 20 μm. In a sensitivity analysis, severe contamination of the aquifer occurs within several decades when the fracture apertures are in the range of 5 to 10 μm, but occurs quickly when the apertures are 20 μm or larger. However, even when the apertures are 25-50 μm, the bulk vertical hydraulic conductivity of the aquitard is still very small ($<10^{-7}$ cm/s). An inference from this study is small, deep, widely spaced vertical fractures may severely diminish the integrity of clayey aquitards without causing much decrease in the bulk vertical hydraulic conductivity. Such fractures can be difficult to detect or locate in field investigations, particularly when only conventional investigative techniques are used. The modeling study by Harrison, Sudicky, and Cherry (1992) suggests the spatial pattern of hydraulic head in fractured aquitards is an important line of evidence in the search for deep, hydraulically active fractures in field monitoring programs.

The discussion presented above concerning contaminant migration through aquitards pertains to non-indurated aquitards (e.g., Quaternary aquitards with considerable clay or silt content), however the same principles apply to lithified aquitards such as shale or mudstone aquitards or carbonate-rock aquitards. These rocks typically have 2-20 percent matrix porosity, which can cause matrix diffusion to be an important factor in solute transport in fractures. Although aquitards can have strong integrity with respect to dissolved contaminants because of the matrix diffusion effect, and in some cases sorption and/or degradation, DNAPLs present a much different problem, discussed in section 4.4.

Experimental Validation of Matrix Diffusion Effects on Solute Transport

After Foster (1975) introduced the concept of solute retardation in fractures due to matrix diffusion, research to assess the importance of matrix diffusion became active in three topic areas: development of improved mathematical models (e.g., Sudicky and Frind 1982, Grisak and Pickens 1980); investigations involving environmental isotopes (e.g., Day 1977, Hendry 1982), and experiments conducted in the laboratory and in the field. Grisak, Pickens, and Cherry (1980) conducted the first major matrix-diffusion experiment using a large cylindrical sample of fractured clayey till taken from from a 5 m (16 ft) deep excavation at the Whiteshell nuclear research site in Manitoba. Breakthrough curves for dissolved inorganic tracers passing through this column, mounted in a large permeameter in the laboratory, were obtained at the column outlet. Outlet sampling showed effects attributed to matrix diffusion. Although this experiment demonstrated the matrix diffusion effects of solutes migrating through fractured clayey till, the procedure used for collecting the sample and the condition of the sample in the permeameter was such that maintenance of the fractures close to the in situ field state was not possible. Jorgensen and Fredericia (1992) developed an improved method for conducting solute transport experiments in large cylindrical samples of clayey material excavated from aquitards. We refer to this as the Danish method. This method enables the in situ stress conditions to be re-imposed in the laboratory. This experimental apparatus has been used in many studies of solute and DNAPL behavior in fractured clayey deposits (e.g., Broholm et al. 1999a; Broholm et al. 1999b; Jorgensen, McKay, and Spliid 1998; Jorgensen et al. 1998; Jorgensen et al. 2002; O'Hara et al. 2000). Contributions of Danish hydrogeologists to understanding contaminant behavior in clayey aquitards have been large since the mid-1990's. Denmark is reliant on groundwater for more than 90 percent of its drinking water and the Danish aquifers are overlain by clayey aquitards.

Although the Danish method for conducting tracer experiments using large cylindrical columns of relatively undisturbed aquitard material has provided considerable information on the behavior of contaminants in fractured materials, corroboration from field experiments is also needed. McKay et al. (1993) and McKay, Cherry, and Gillham (1993) conducted a field tracer experiment using bromide and chloride as solute tracers to assess matrix-diffusion effects in a natural, densely fractured, clayey glaciolacustrine deposit near Sarnia, Ontario. They imposed a steady horizontal groundwater flow in the surficial fractured zone of the aquitard in which the colloidal and solute tracers were transported horizontally from one gravel-filled trench to another spread 2-4 m apart and used colloid-size particles to measure the groundwater flow rate through the fracture network. Solute tracers are subject to matrix diffusion effects but colloids are not. The colloids traveled rapidly (< 2 days) from one trench to another and the solute tracers much more slowly (several weeks), consistent with modeling showing strong matrix diffusion effects on the solutes. The rapid migration of colloid tracers was an important observation demonstrating the relevance of the calculation procedure (presented in Chapter 3) for average linear groundwater velocity in which the Darcy flux is divided by the bulk fracture porosity. This type of calculation typically provides a large groundwater velocity in fractured aquitards, on the order of meters per day, which is the velocity of the colloid tracer in the McKay et al. tracer experiment.

DNAPL ENTRY AND FLOW IN AQUITARDS

Aquitards with a strong capability to protect underlying aquifers from dissolved contamination should not necessarily be expected to provide strong protection in situations where contaminants moved in the DNAPL state. The large difference between the propensity for solutes and DNAPL to migrate through aquitards is caused by differences in the relative strength of matrix diffusion effects compared to fluid flux (i.e., advection) in the fracture. In the modeling assessment of TCE migration by Harrison, Sudicky, and Cherry (1992), fully penetrating fractures with apertures as small as 5 μm did not result in aquifer impacts over practical time scales (e.g., centuries) because the diffusion-driven retardation effect was extremely strong. However, when DNAPL with properties typical of chlorinated solvent DNAPLs enters and then flows in fractures, matrix diffusion has little or no influence on the flow rate (Kueper and McWhorter 1991). Also, because of the large liquid density and small interfacial tension of common DNAPLs such as chlorinated solvents, substantial DNAPL accumulation on top of an aquitard causes a strong propensity for the DNAPL to enter very small fractures, even those with apertures smaller than 10 μm (Kueper and McWhorter 1991, McWhorter and Kueper 1996). Therefore, compared to the hypothetical case simulated by Harrison, Sudicky, and Cherry (1992) where the TCE entered the fractured aquitard in dissolved phase, much different results would be expected if the TCE entered the system as a dense immiscible-phase liquid (i.e., as DNAPL). Using the Danish large column method, O'Hara et al. (2000) conducted a laboratory experiment involving a meter long cylindrical sample of glaciolacustrine clay with vertical fractures in which free-phase TCE DNAPL was ponded on top of the column. The DNAPL entered fractures with apertures in the 5 to 15 μm size range and flowed through the column in less than two weeks. Model simulations indicate dissolved TCE would take many years or longer for groundwater transport through this column, in contrast to the rapid DNAPL flow. Although the clay column had fractures allowing rapid DNAPL flow, they were so small the measured bulk vertical hydraulic conductivity of the column did not provide evidence of fractures (i.e., only slightly larger column K than the matrix K). Field experiments involving small volume releases (3-5 L) of TCE and PCE DNAPLs conducted in the fractured zone of the Sarnia clay showed the relative mobilities of these two solvent DNAPLs within the fracture network inversely related to their diffusion coefficients and viscosities (Kirkpatrick 1998).

Anecdotal evidence suggests chlorinated solvent DNAPLs, such as TCE and PCE, have migrated through many aquitards to cause contamination of underlying aquifers. However published information on this topic is sparse. One indirect line of evidence suggesting DNAPLs have migrated through many aquitards is volatile organic compounds (VOCs) such as TCE, PCE and dichloromethane (DCM) commonly occur in aquifers formerly believed to be well protected by overlying aquitards. In some cases, these VOCs occur without the presence of other contaminant types, such as chloride or nitrate or atmosphere-derived tritium, originating only in aqueous form and never as DNAPL. Mclelwain, Jackman, and Beukman (1989) describe a site in southwestern Ontario, Canada where DNAPLs, primarily PCBs, TCE and trichlorobenzene, leaked from a drum storage facility, and moved downward through a 6-7 m thick surficial clay aquitard of glaciolacustrine origin to cause severe and persistent contamination of the underlying bedrock aquifer. In the investigation conducted after this contamination was discovered, fully penetrating fractures were mapped in an excavation dug through the aquitard to the bedrock surface (Wills et al. 1992). The fractures probably originated long ago due to contraction of the clay caused by cycles of wetting and drying and freezing and thawing. The drum storage facility was established in 1978 without hydrogeologic studies and the bedrock contamination was discovered in 1985. The fractured nature of the aquitard could have easily been recognized by excavation and other methods prior to selection of the site for drum storage. Another example of a clayey aquitard that allowed passage of DNAPL through and into an underlying aquifer was reported by PPG Industries (1995) for a site in southwestern Louisiana. The DNAPL originated from a chemical manufacturing facility from which large quantities of chlorinated solvent DNAPL were released into the ground. The aquitard extends from ground surface to a depth of 50 m (164 ft). The aquitard originated as mud in overbank deposits from a large river. An intensive site investigation showed the presence of ethylene dichloride (EDC), TCE and PCE in the upper zone of a regional sand aquifer system (the Chicot Aquifer) downward DNAPL movement via fractures and / or root holes through the full 50 m (164 ft) aquitard thickness. These preferential pathways are consistent with the overbank depositional origin of the aquitard. The presence of fractures was evident in core samples and indirectly from the presence of atmospheric-derived tritium deep in the aquitard. Hanor (1993) described a study of a landfill site on a silty and clayey aquitard of fluvial-deltaic origin in southeastern Louisiana where the bulk vertical hydraulic conductivity of the aquitard was up to 4 orders-of-magnitude larger than values obtained from laboratory measurements on core samples. He concluded the large hydraulic conductivity was caused by fractures formed during periods of subareal weathering. Contaminant migration was not investigated in this study, however in the context of engineered landfills, Hanor states "There is no reason to believe that the sediments at this site or sites in similar geologic settings in the Gulf Coast will provide an adequate barrier to contaminant migration in the event of failure of an engineered containment system".

Sabourin (1989) provides a field example where DNAPL, in this case a combination of two miscible wood-preservative chemicals, creosote and pentachlorophenol (PCP), flowed downward through vertical fractures in the surficial glaciolacustrine aquitard in Winnipeg, Manitoba. The creosote (black, oil-like liquid) was observed in vertical cores down to the bottom of the brown, weathered part of the aquitard at 8.5 m and indirect lines of evidence indicated it also penetrated through the underlying 3 m thick zone of gray, unweathered clay. Dissolved polycyclic aromatic hydrocarbons (PAHs) and PCP were detected across the entire thickness of the gray clay down to the top of the till. A laboratory study of the diffusion of two PAHs (naphthalene and anthracene) and PCP through water-saturated gray clay was conducted using undisturbed core samples in diffusion cells. The diffusion of PAHs and PCP was very slow because of adsorption. Using the hypothesis of the gray clay being unfractured, the dissolved wood-preservative compounds would migrate downward, driven by molecular diffusion with strong retardation by sorption. Modeling showed this hypothesis unable to explain the presence of PAHs and PCP observed throughout the gray clay because the modeled diffusion distances are too short. The hypothesis of gray clay with vertical fractures allowing downward advection of the dissolved compounds also provided a poor match to the field data.

The only plausible explanation for the contaminant distribution in the bottom part of the aquitard is DNAPL flow downward through vertical fractures (spaced ≈ 1m apart) in the gray clay with horizontal diffusion of dissolved constituents from the DNAPL into the clay matrix blocks between fractures. This conceptual model provides high probability of contaminant detections in vertical cores consistent with the field observations.

The creosote site in Winnipeg investigated by Sabourin (1989) is close to the site in Winnipeg studied by Day (1977) where tritium penetrated no deeper than 5 m (16 ft). At the site studied by Sabourin tritium was also found no deeper than 5 m (16 ft) depth, however the creosote-derived compounds occurred down to the bottom of the aquitard at 11-12 m (36-39 ft). These studies of the Winnipeg aquitard demonstrate deep DNAPL penetration in small fractures that did not allow deep penetration of dissolved constituents such as ^{3}H and ^{18}O that were strongly retarded by matrix diffusion.

Although deep penetration of DNAPL in fractured aquitards has been documented at several sites, intensive field studies also show some aquitards capable of preventing DNAPL penetration and others allowing partial but not full penetration. Thus, some aquitards have excellent integrity even when DNAPLs are the contaminant source. Roberts, Cherry, and Schwartz (1982) and Schwartz, Cherry, and Roberts (1982) provided the first published descriptions of DNAPL penetration into but not through a thick layered, silty and clayey Quaternary aquitard beneath Regina, Saskatchewan. At this site, the PCB oil (i.e., DNAPL) flowed down vertical fractures in the surficial aquitard unit composed of 9 m (29 ft) of glaciolacustrine clay. The DNAPL penetrated to the bottom of this clay unit, above the water table (i.e., the vadose zone), but did not penetrate into the underlying silt layer of the aquitard. A lack of continuity of the fractures in the clay with those in the silt unit may have prevented deeper flow of the DNAPL, or perhaps the fracture apertures in the silt were too small to allow DNAPL entry.

Morrison, Parker, and Cherry (1998) and Parker (1996) describe a field example where DNAPL penetrated into a thin surficial, clayey aquitard but did not go to the bottom of the aquitard. The aquitard allowed DNAPL entry via vertical fractures connected to thin horizontal sand seams at depth within the aquitard. This field site is situated at Canadian Forces Bases Borden, Ontario, where a PCE DNAPL release experiment was initiated in 1991, when 771 L (204 gal) of free-product PCE were infiltrated into a 4 m thick surficial sand aquifer underlain by the clayey aquitard (Brewster et al. 1995). The aquitard is unweathered and, visually, appeared unfractured. However, after the DNAPL was found in the aquitard, a comprehensive study using several other methods, including ^{3}H and a pumping test, clearly indicated the presence of vertical pathways (i.e., fractures) in the upper part of the aquitard. A detailed profile of hydraulic head versus depth showed the bottommost 0.5 m of this 8 m (25 ft) thick aquitard has a much lower bulk vertical hydraulic conductivity than the shallower part of the aquitard (Figure 3.10). Therefore, the bottom zone, which may be unfractured, provided nearly all of the resistance to downward groundwater flow through the aquitard. The sand seams allowed lateral DNAPL flow and DNAPL storage, which likely helped to dissipate the downward DNAPL driving force. This case study demonstrates the part of an aquitard that provides nearly all of the aquitard integrity may be only a thin zone within the overall aquitard. Therefore, in investigations of aquitard integrity, the strategy should ensure identification of these zones so that they can be investigated appropriately. Evidence of thin, very low K layers within "aquitard" formations has been obtained at several sites in both non-indurated and indurated aquitards (e.g., Eaton and Bradbury 2003) where detailed depth-discrete hydraulic head profiles were measured.

Parker, Cherry, and Chapman (2004) describe a contaminated site investigation in Connecticut where free-product TCE DNAPL formed a pool at the bottom of a 10 m (33 ft) thick sand aquifer. The DNAPL did not penetrate below the aquifer bottom because of the presence of an underlying 18 m (59 ft) thick silty and clayey glaciolacustrine aquitard. Analysis of cores from the top of the aquitard shows TCE occurrence in the upper part of the aquitard was caused by downward diffusion of dissolved TCE and not DNAPL, except possibly in the uppermost 20 cm. Mathematical modeling of TCE diffusion indicates TCE would take many hundreds of years or longer to diffuse to the bottom of the aquitard. The lack of DNAPL penetration into the aquitard was attributed to an absence of open fractures; this aquitard is composed of soft silt and clay with varved layering. Possibly, either the sand that forms the surficial aquifer was deposited on top of the aquitard without an intervening episode allowing aquitard dessication, or the soft, plastic nature of the sediment caused closure of all fractures that may have formed prior to the deposition of the surficial sand. Although it is intuitively reasonable to expect fractures opened in a soft, plastic, clayey deposit during one geologic episode to become closed in a later geologic episode when the fractured clay deposit acquires increased confining load due to more recent deposits overlying, the literature does not contain quantitative assessments using geomechanics principles to define the stress conditions necessary for complete fracture closure.

EFFECTS OF DNAPL ON AQUITARD PERMEABILITY

In the early 1980s, laboratory research suggested that some types of industrial organic chemicals, including chlorinated solvents, could interact chemically with clay-rich materials to cause shrinkage and cracking, thereby producing causing a large permeability increase with respect to water and DNAPL (Brown, Thomas, and Green 1984; Green, Lee, and Jones 1981). This research was sponsored in part by the U.S.EPA and consequently the results had much influence, even though the experiments were limited in rigor and scope and caused controversy in the scientific community. Based on these experiments, some groundwater scientists and engineers held the belief chlorinated solvent accumulations on top of clayey deposits would result in deep DNAPL penetration due to physio-chemical alteration of the clays (i.e., chemically inducted fractures) regardless of whether or not fractures existed prior to arrival of the DNAPL on the clay layer. The controversy associated with this topic prompted several follow-up laboratory studies.

The published results of studies of the effects of organic chemicals on clays show contradications. With few exceptions, the literature is based on laboratory experiments. The study of clay/organic chemical interactions spans many disciplines, including colloid and surface science, geotechnical engineering, clay mineralogy, organic chemistry and hydrogeology; the interdisciplinary nature of this topic has likely contributed to the early lack of agreement in the literature. Although few of the laboratory studies used chlorinated solvents, many used other organic chemicals that in some cases provided results relevant to chlorinated solvent-clay interactions.

Middleton and Cherry (1996) conducted laboratory experiments and a literature review to assess the "chemically induced fractures" hypothesis. Consideration of the relevant factors such as water miscibility of organic liquids as well as the details of the published laboratory experiments provided a basis for resolving many of the contradictions in the literature. In hydrated (i.e., wet) clayey deposits, they concluded chlorinated solvents and other NAPLs apparently do not cause permeability increases, likely because of the exclusion of the organic liquid from the clay-mineral double layers. However, in laboratory experiments when NAPL solvents were forced under large applied pressure through water-saturated clay samples, permeability increases were observed. These increases are attributed to conditions by the experimental apparatus causing the mechanical formation of fractures due to large fluid gradients and low lateral confining stress on the samples. The gradients needed in laboratory experiments to cause permeability increase are much larger than can realistically occur in the field. Exposure of the natural clayey materials to these organic compounds in aqueous solutions does not influence permeability because the maximum aqueous concentrations allowed by solubility are too small to influence the clay double layers. Parker (1996) reported on long-term field experiments in which chlorinated solvent DNAPLs were placed as free-product columns at the bottom of uncased boreholes in a thick clayey aquitard near Sarnia, Ontario. Cores collected from below the DNAPL after 6.5 years showed no DNAPL penetration. As expected, penetration of dissolved phase solvents was governed by diffusion and sorption. The contaminated site studied by Parker, Cherry, and Chapman (2004) described above provides additional evidence for lack of DNAPL penetration into natural unfractured, water-saturated clayey aquitards.

At the contaminated site in Smithville, Ontario, dense oils containing PCBs, TCE and chlorobenzenes moved through a natural clay deposit and entered an underlying fractured limestone aquifer (McIelwain, Jackman, and Beukman 1989). Initially, chemical interactions between the dense oils and clay were suspected to have facilitated movement through the clay. However, the DNAPLs were later concluded to have moved downward though pre-existing fractures in the clay deposit. At a site in Wilsonville, Illinois, organic chemicals including chlorinated solvents, pesticides, and aromatic compounds migrated through a clayey glacial till at rates 100 to 1000 times greater than had been predicted based on the assumption of initially unfractured deposits (Griffin et al. 1984, Herzog et al. 1989); however, the primary reason for the rapid migration was later discovered to be the presence of pre-existing fractures in the till. Nevertheless, given the prevailing view at the time the investigators erroneously concluded organic-clay interactions may have caused permeabilities larger than those attributable to the natural fractures.

ENVIRONMENTAL ISOTOPES AND NATURAL CHLORIDE AS TRACERS FOR MIGRATION PATHWAYS AND RATES

Assessments of aquitard integrity commonly benefit from determining the distribution of particular natural dissolved constituents and isotopes in the aquitard porewater. These natural constituents serve as tracers of the groundwater flow system and can provide identification of potential contaminant migration pathways and mechanisms. The groundwater flux is sufficiently small in many aquitards for the aquitard water and the natural or anthropogenic dissolved constituents within to be sufficiently old for the distributions to be influenced mostly or exclusively by diffusion. The natural dissolved constituents commonly provide insight about aquitard integrity because they have been present long enough, commonly one geologic time scale, for their spatial distributions to provide reliable estimates of solute migration rates. These dissolved natural constituents serve as analogs for dissolved contaminants that may arrive at or in the aquitard in the future. Chloride is the most common natural major ionic constituent used in aquitard studies and tritium (^{3}H), oxygen-18 (^{18}O) and deuterium (^{2}H) are the most commonly used isotopic constituents. Atmospheric-derived ^{3}H is included here in the category of natural isotopic tracers even though much of it is present in the atmosphere due to human activities (i.e., atmospheric testing of nuclear devices in the 1950's – early 1960's). Natural carbon-14 (^{14}C) is also useful in some situations. Also, in some situations the chloride in aquitards is anthropogenic. The spatial distributions of the diagnostic constituents are determined by analyses of groundwater obtained from piezometers or extracted from cores. Trends in the spatial distribution of these constituents sometimes indicate the presence of fractures and their potential influence on solute transport. Insight into aquitard integrity is obtained by fitting transport simulations over the relevant time period (e.g., decades, centuries or millennia) to the concentration distributions obtained from the field. In this approach mathematical models calibrated to distributions of chloride and/or the environmental isotopes are used to simulate contaminant migration in the aquitards.

Clark and Fritz (1997) and Mazor (2003) describe the general uses of environmental isotopes in hydrogeology. The stable isotopes, ^{18}O and ^{2}H, (measured as ratios of $^{18}O/^{16}O$ and $^{2}H/^{1}H$ relative to a standardized sample of sea water) commonly have signatures in aquitard porewaters indicative of past climatic conditions or indicative of water sources different from the present-day (e.g., modern) groundwater. For example, stable isotope signatures representing the colder conditions existing during Pleistocene time are commonly found in thick clayey aquitards in the glaciated part of North America (e.g., Farvolden and Cherry 1988; Simpkins and Bradbury 1992; Remenda, van der Kamp, and Cherry 1996; Hendry and Wassenaar 1999).

Tritium is radioactive hydrogen with a decay half-life of 12.3 years and occurs as tritiated water molecules ($^{3}H{\cdot}HO$). Global precipitation became contaminated with ^{3}H from atmospheric tests of nuclear devices beginning in 1952, and atmospheric ^{3}H concentrations rose to a maximum in 1963 at which time the ban on above-ground nuclear tests went into effect. The ^{3}H concentrations in global precipitation declined rapidly in the next few years followed by a more gradual decline thereafter. ^{3}H in precipitation is documented by monitoring stations around the globe. The general trends are similar from station to station, in the northern hemisphere with lower concentrations from north to south (Clark and Fritz 1997). Several laboratories in North America and Europe analyze for ^{3}H in water samples to the extremely low detection limits (0.1 or 0.01 Tritium Units (TU)) generally needed for use in hydrogeological investigations. These detection limits are two to three orders of magnitude below the ^{3}H concentration in present-day precipitation and also far below the ^{3}H peak of 1963 adjusted for ratioactive decay. Analyses using the lowest-detection limit are generally best-suited for aquitard studies. The literature concerning ^{3}H in groundwater pertains mostly to granular aquifers, where ^{3}H is commonly used to determine groundwater travel times or groundwater age. The presence of ^{3}H in aquifers identifies recharge water that entered the aquifers since the early 1950's. In some cases, the 1963 ^{3}H peak is found in aquifers, providing a precise time line. However in aquitards, ^{3}H is not a groundwater travel time or age indicator because, in unfractured aquitards, diffusion rather than advection (i.e., groundwater flow) commonly dominates, and in fractured aquitards, the matrix diffusion effect causes ^{3}H retardation relative to groundwater velocity in fractures.

Day's (1977) study of the Winnipeg aquitard was the first to indicate usefulness of ^{3}H, ^{18}O, and ^{2}H in aquitard investigations. Day (1977) observed a strong downward hydraulic gradient in the aquitard directed from the water table to the underlying regional aquifer, which has been pumped strongly for more than a century. ^{3}H indicating presence of some modern water molecules (> 1950) only in the upper 5 m of the 13 m thick aquitard. ^{18}O indicates presence of Pleistocene age water molecules in the middle of the aquitard, even though vertical fractures occur from ground surface to the bottom of the aquitard. Day (1977) accounted for the spatial patterns of ^{3}H and ^{18}O by using the combined influences of active advection in the fractures and diffusion-driven mass transfer between the fractures and the clay matrix. The absence of ^{3}H at greater depth in the aquitard provided evidence that although the aquitard has fractures throughout and a strong downward hydraulic gradient, the rate of downward migration of solutes is very slow relative to the calculated average linear groundwater velocity, which is on the order of meters per day. The existence of Pleistocene-age ^{18}O in the aquitard after nearly a century of strong downward flow is further evidence of aquitard integrity, but only in the context of solute transport. Wood preservative DNAPL was found much deeper than the deepest ^{3}H. Hendry (1982) observed ^{3}H throughout the full thickness of a clayey till aquitard in southern Alberta, Canada indicating fracture pathways. In the past two decades many other investigators have used these isotopes in aquitard studies. In the Sarnia, Ontario area, Ruland, Cherry, and Feenstra (1991) observed ubiquitous ^{3}H in the brown weathered clay within 3-5 m of ground surface and decreasing ^{3}H in the upper several meters of the underlying thick gray unweathered clay confirming the presence of hydraulically active fractures penetrating into the unweathered part of the aquitard.

Farah, Parker, and Cherry (2005) used a numerical model (FRACTRAN) to assess the potential for atmospheric derived ^{3}H in surficial clayey aquitards to serve as a tool to identify the presence of hydraulically active, widely spaced vertical fractures at depth. They simulated the ^{3}H distribution in a hypothetical 15 m (49 ft) thick aquitard with downward groundwater flow through vertical fractures. Fracture abundance, connectivity, aperture, depth, spacing, and orientation, in addition to the hydraulic gradient, controlled ^{3}H distribution. However, the method was found to be insensitive to very small fractures ($\leq$ 10 μm, 3.9×10^{-4} in) using the lowest available detection limit of 0.1 TU; therefore, the maximum depth of detectable ^{3}H is commonly much shallower than the maximum depth of active fractures. Consequently, substantial concentrations of ^{3}H deep in clayey aquitards may indicate presence of active larger fractures ($\geq$ 20 μm) or other types of preferential pathways. The modeling indicates the ^{3}H signature in surficial aquitards of the type occurring in many regions is a useful diagnostic tool in assessment of aquitard integrity even though radioactive decay has greatly reduced ^{3}H concentrations since the peak concentrations occurred in the 1960's. Farah, Parker, and Cherry (2003) conducted FRACTRAN simulations of the same idealized aquitard-aquifer situation indicated above and examined the steady-state hydraulic head distribution in combination with the ^{3}H distribution. They concluded vertical and horizontal profiles of head used in combination with the ^{3}H distribution provide the best prospects for identifying the presence and nature of deep, widely spaced fractures relevant to aquitard integrity. However, for the profiles to be strongly diagnostic, the vertical or horizontal distance between data points must be smaller than distances used in common practice in aquitard monitoring.

For situations in which the integrity of a laterally extensive, surficial clayey aquitard with downward groundwater flow is being assessed, the ^{3}H distribution can be a strong indicator of the capability of the aquitard to protect the underlying aquifer from migration of dissolved contaminants. If the field determination at a site shows detectable ^{3}H has not yet penetrated to the bottom of the aquitard and if the hydraulic gradient has been downward for at least a few decades, then it can be expected that no other species of dissolved 'contaminant' that entered the subsurface since the 1950's will also likely not have fully penetrated the aquitard. However, for any particular aquitard-aquifer system, this general expectation (hypothesis) should be evaluated in the context of the site-specific factors with consideration of the maximum contaminant level (MCL) compared to the source concentration (i.e., C/C_o) for the contaminants of concern. However, the lack of deep ^{3}H penetration in an aquitard should not be taken as evidence for good aquitard integrity with respect to DNAPLs or particulates such as viruses or colloid-facilitated transport because matrix diffusion does not diminish the mobility of these contaminant types.

The approach described above for use of ^3H in aquitard investigations is based on determination of the ^3H distribution within aquitards. However, another useful approach involves determination of ^3H presence or absence in aquifers beneath aquitards. ^3H presence in the underlying aquifer indicates the existence of rapid flow paths extending from land surface to the location of the observed ^3H. These rapid flow paths may exist in the aquitard, or exist entirely within the underlying aquifer if the aquifer is connected laterally to recharge zones. In an investigation of a deep bedrock aquitard in southeastern Wisconsin, Eaton (2002) found abundant ^3H in the overlying aquifer, and no detectable tritium in the underlying aquifer, which suggests favorable aquitard integrity in that area.

The regional surficial aquitard in the Sarnia area of southwestern Ontario, for which the general hydrogeology is described by Husain, Cherry, and Frape (2004), has been subjected to many types of studies using environmental istopes relevant to aquitard integrity. The surficial weathered, fractured zone of this aquitard is 3-5 m thick (e.g., McKay and Fredericia 1995). Desaulniers (1980), Desaulniers, Cherry, and Fritz (1981) and Desaulniers et al. (1986) found the thick sequence of unweathered till beneath this surficial zone exhibits isotope (^{18}O, ^2H, ^{14}C, ^{37}Cl) versus depth and natural chloride versus depth profiles indicative of old groundwater, probably in the range of 9,500 – 14,000 years b.p., persisting near the bottom of the aquitard. Desaulniers (1986) and Husain et al. (1998) showed isotope and chloride profiles, with only rare exceptions, closely match simulations of one-dimensional solute migration based solely on molecular diffusion. These isotopic indications of diffusion control in the unweathered part of the aquitard below the surficial weathered zone are consistent with contaminant distributions below waste disposal pits. For example, Johnson, Cherry, and Pankow (1989) obtained chloride profiles in the unweathered clay beneath a hazardous waste disposal pit in the Sarnia area indicative of downward migration under control diffusion, as did Crooks and Quigley (1984) for sodium and chloride at one municipal landfill in the same area. Yanful and Quigley (1990) measured profiles of ^{18}O, ^2H and ^3H in the aquitard beneath this landfill that also closely matched the diffusion model. The landfill leachate had different isotropic signatures than the background aquitard pore water. A comprehensive modeling analysis of downward diffusion of solutes from landfills in Sarnia aquitard is provided by Quigley and Rowe (1986).

A general conclusion drawn from the many studies of the regional clayey aquitard in the Sarnia area of southwestern Ontario is the thick zone of unweathered clay comprising the bottom part of the aquitard can be expected to provide long-term restriction on downward contaminant migration, with diffusion times for breakthrough into the underlying aquifer on the order of many hundreds to thousands of years. This conclusion is based on data obtained since the field studies began in the 1970's, which consistently indicate hydraulically-active fractures do not penetrate to depths greater than about 15 to 20 m. This degree of aquitard integrity is not expected where the total aquitard thickness is less than about 25 m and may not occur at locations where regional structural features in the underlying bedrock have influence on fracture occurrence in the aquitard.

Another regional clayey aquitard subjected to many investigations relevant to aquitard integrity occurs in the St. Lawrence Lowland in the Province of Quebec. During Pleistocene time, an inland sea (the Champlain Sea) connected to the Atlantic Ocean occupied the St. Lawrence Lowland. The sea was a mixture of fresh water from melting glaciers and ocean water. Extensive deposits of clay formed with maximum thickness of 50 m. In the Montreal area, Lafleur, Giroux, and Huot (1987) found the hydraulic conductivity of the weathered zone (0-5 m) is 2 to 4 orders of magnitude larger, due to fractures, than the underlying unweathered clay ($K \approx 10^{-8}$ cm/s, $K \sim 3x10^{-10}$ ft/s) beneath. This conclusion is derived from packer-injection tests in reamed auger holes. The packer-injection assembly lowered down the boreholes has a cutting head that enlarges the borehole by slicing away the smear zone on the borehole wall caused by the augering process. With the smear zone removed, the fractures are exposed and the injection tests yield K values much higher than would otherwise be the case. This field study, along with a field study by D'Astous et al. (1989), demonstrates the importance of installing piezometers in a manner that minimizes detrimental effects of drilling.

Quigley et al. (1983) and Desaulniers and Cherry (1989) conducted studies of the Champlain Sea clay with emphasis on the unweathered clay below the surficial weathered zone. They concluded molecular diffusion is a major factor in migration of natural major ions and isotopes in the clay. High major-ion concentrations (e.g., brackish water) originated in the clay pore water when the clays were deposited in the Champlain Sea about 10,000 – 11,000 years ago. When the Sea drained and the landscape was exposed to fresh water (rain and snow), the verticall concentration gradient caused these ions to diffuse upward against the downward hydraulic gradient, creating the gradual depletion in major-ion concentrations at shallow depths existing today. Desaulniers and Cherry (1988) used a one-dimensional diffusion model to produce simulations closely matching the field concentration profiles. Quigley et al. (1983) also obtained close model fits based on salt and isotope diffusion with no advection.

Two methods have been commonly used to obtain groundwater samples for ^{18}O and ^{2}H analyses: squeezing water from core samples and sampling standpipe-type piezometers (e.g., Remenda 1993, Husain 1996). These two methods give similar, but generally not identifical, results. Kelln, Wassenaar, and Hendry (2001) provide a comparison of these two methods with three methods involving core samples: pore-water extraction using a centrifuge, azeotropic distillation and a direct soil-water equilibrium technique using carbon dioxide. They reported that the equilibrium method yielded sufficiently accurate and reproducible ^{18}O results with the advantage of eliminating the need for labor intensive complete extraction of the water from the geologic medium. Small differences in the ^{18}O values between piezometer water and equilibrated, squeezed and centrifuged samples suggested that each method collected different fractions of the clay-water reservoir.

As indicated above, the distribution of major ions such as Cl- and Na+ in groundwater in clayey aquitards may provide useful insights about solute migration, however, it is important that the chemical analyses have minimal bias caused by the sampling methods. Remenda and van der Kamp (1997) demonstrated strong adverse effects of reactive bentonite seals of the type commonly used in piezometer seals in aquitards. Piezometers in aquitards are much more prone to such influences than those in aquifers because of extremely slow inflow of water to the standpipe reservoir. Also, Wassenaar and Hendry (1999) point out that inadvertent introduction of microbial contamination or oxygenation of the intake screen during piezometer construction and subsampling could stimulate in-piezometer microbial mediated reactions such as sulfate reduction or affect redox species (i.e., Fe^{2+}, H_2S, As^{3+}), thereby resulting in misleading or changing groundwater chemistry results not representative of the aquitard pore water. Wassenaar and Hendry (1999) implemented and evaluated an aquitard piezometer installation and groundwater sampling strategy and showed that the use of an inert gas pocket in the piezometer construction can be used to delay seal contamination for at least three years and avoid oxygenation and disturbance of downhole redox conditions.

Carbon-14 (^{14}C) is another isotope used to examine natural solute migration in clayey aquitards. In granular aquifers (e.g., sand or gravel) or other types of aquifers where matrix diffusion effects are minimal, ^{14}C can, in some circumstances, be used to determine groundwater age (Clark and Fritz 1997), however in aquitards where diffusion exerts control of solute migration, ^{14}C is not a groundwater age-dating tool; rather the ^{14}C distribution of dissolved carbon can in some cases be used to establish existence of long-term diffusion control. For example, Wassenaar and Hendry (2000) used ^{14}C and ^{13}C measurements of dissolved inorganic carbon to demonstrate diffusion controlled conditions in a thick clay-till aquitard in southern Saskatchewan. Hendry and Wassenaar (2005) gained additional insight into solute behavior in this aquitard using the ^{14}C distribution of natural dissolved organic carbon.

SIMULATING FLOW AND TRANSPORT THROUGH FRACTURED AQUITARDS

In the analysis of flow and transport through fractured aquitards, an important issue pertains to whether the fractured medium can be usefully simulated as an equivalent porous medium (EPM) or whether preferential transport occurs through fractures in a manner that renders EPM simulations useless or misleading. Fractured porous geologic media can be conceptualized as a dual-porosity system containing porous matrix blocks between fractures. Many mathematical models exist for representing solute transport in fractured porous media in which the matrix blocks have sufficient porosity for matrix diffusion to be important. There are two categories of models: discrete-fracture and dual-porosity. Discrete-fracture models represent the fractures as discrete planar entities, either single parallel plates, multiple parallel plates or a network of interconnected fractures. Two porosity domains exist in dual porosity models, one with active flow and the other with no flow or minimal flow providing a reservoir for mass transfer and storage while interacting with the active flow domain. Other recently developed methods incorporate the effect of discrete fractures into porous media models such as MODFLOW (Langevin 2003; Rayne, Bradbury, and Muldoon 2001). The criteria that must be met for the EPM approach to be valid for representing solute transport in fractured media are described by van der Kamp (1992).

The EPM approach for representing solute migration is most appropriate for those situations where the fracture spacing is relatively small and the matrix porosity relatively large (i.e., large effective diffusion coefficients for the solutes) such that the diffusion-driven mass transfer from the fractures into the low-permeability matrix blocks between fractures is strong relative to the advective transport along the fractures. Thus, the matrix blocks are quickly "filled" with contaminant mass as the plume front advances. Pankow et al. (1986); Mckay, Gillham, and Cherry (1993); McKay, Balfour, and Cherry (1998) describe field examples where the EMP approach was suitable for analysis of solute migration in fractured clayey aquitards. In these cases the solutes were migrating laterally at shallow depth in the intensely fractured clayey aquitards. Pankow et al. (1986) also describe a field case (fractured shale) where the EPM approach was not suitable.

Fractures and fracture networks are complex in the field and therefore difficult to represent realistically in mathematical models. In fracture networks, individual fractures have variable apertures and there is much variability of fracture length. Some fractures are well connected to other fractures, and others are not, and some fractures have weathering effects or sediment infill. Nevertheless, mathematical models can play useful roles in assessing fractured aquitard integrity. Several computer codes that do not use the EPM approach have become useful for modeling groundwater flow and solute transport in discretely fractured media:

CRAFLUSH, developed by Sudicky and Frind (1982), FRACTRAN, developed by Sudicky and McLaren (Sudicky and McLaren 1992), FRACMAN/MAFIC, developed by Golder Associates (Dershowitz et al. 1993) and FRAC3DVS, Therrien and Sudicky (1996).

CRAFLUSH is a one-dimensional fracture flow and transport code that simulates flow in parallel fractures with matrix diffusion. Toran (2000) developed a companion code, CRAFIT, that automates trial and error fitting of breakthrough curves generated by CRAFLUSH. These codes are useful for first approximations of contaminant migration through material thought to contain parallel fractures.

In FRACTRAN, a network of orthogonal fractures is generated according to specified statistical (Gaussian) distributions of fracture aperture spacing and length. The fractures can have variable aperture, in which case they must all have the same mean aperture, or each fracture has a single aperture, but the aperture can vary between fractures. Because the fracture network has assigned statistical properties, each fracture connects to one or more fractures depending on fracture length and spacing. This model has restrictions in its applicability to some field sites because it is limited to two-dimensions and permits only orthogonal fracture arrangement. Nevertheless, FRACTRAN is a major advance over other models that represent only uniform parallel plate fractures or that represent the system as dual-porosity domains without direct incorporation of fracture network features. Harrison, Sudicky, and Cherry (1992) provide an example of the use of FRACTRAN to simulate dissolved contaminant migration through an idealized fractured aquitard.

The FRACMAN/MAFIC code (Dershowitz et al. 1993) combines a detailed fracture-network generation routine with a finite-element numerical solver. FRACMAN generates three-dimensional fracture networks based on measurable statistical properties, such as fracture length, orientation, and density. A unique feature is a powerful set of stochastic modeling and sampling routines to help determine the best-fit realizations of statistical fractures for a given set of field measurements of fracture and hydraulic properties. The MAFIC solver accommodates three-dimensional advection, dispersion, matrix diffusion, sorption, decay, and biodegradation in fracture networks, however its treatment of the effects of diffusion may not be adequate for fractured clayey or lithified aquitards. STRATAFRAC (http://fracman.golder.com/software/stratafrac.asp), a recently developed companion code, translates discrete facture network properties into an equivalent porous media framework that can be solved using porous media models.

The FRAC3DVS code (Therrien and Sudicky 1996) combines a representation of a porous matrix with a set of three dimensional planes representing fractures. It allows a fully three-dimensional representation of fracture network connectivity, solute advection and diffusion, as well as variably saturated flow. However, both FRACMAN-MAFIC and FRAC3DVS are computationally intensive. For this reason simulating actual solute behavior in three dimensions in real fracture networks on more than a very limited scale is currently not practical. Future developments in computational power are expected to make these codes more suitable for application to real sites in which case acquisition of relevant three-dimensional field data will become the limiting factor. At the present time combinations of existing models including EPM and discrete fracture models can be used in different ways at different scales to help resolve questions about groundwater flow and contaminant migration. Different models can address different questions and therefore complement one another.

Numerical models of groundwater flow in which all of the hydrostratigraphic units, including the aquitards, are represented as equivalent porous media, are often used to extrapolate aquitard and aquifer characteristics over large geographic areas through model calibration. For example, Krohelski et al. (2000) estimated the vertical hydraulic conductivity (K_v) of the Eau Claire shale, a regional aquitard in central Wisconsin, as 0.0002 m/d (0.0006 ft/day) by calibration of a regional groundwater flow model also using a regional numerical model. Feinstein et al. (2004) estimated the K_v of the Maquoketa Formation, an aquitard comprised of Ordovician shale aquitard in eastern Wisconsin, as 1.5×10^{-6} m/d (5×10^{-6} ft/d). This value incorporates the effect of numerous deep wells open through the aquitard that allow cross flow. The estimation was robust because the regional model was calibrated simultaneously to two different populations of head and flux targets taken during stressed and unstressed periods (Hart and Bradbury, N.d.). However, as with leakage coefficients derived from pumping tests, such model values are lumped parameters that cannot reflect the spatial heterogeneity including preferential flow avenues, such as windows and fractures present in most aquitards. Therefore this type of information has limited utility for assessing solute transport times through aquitards.

The FRACTRAN and FRAC3DVS numerical models are discrete-fracture, matrix diffusion models (DFMD models) in that they represent flow and solute transport in fractured networks in which each fracture is incorporated individually into the model and diffusion-reactions and flow are represented in each of the low-permeability matrix blocks between the fractures. Although these two models have existed for many years (Sudicky and McLaren 1992, Therrien and Sudicky 1996), they have rarely been used to represent or assess actual field situations. The more common use of these models is examination of fractors influencing solute transport through sensitivity analyses of idealized hydrogeologic situations (Harrison, Sudicky, and Cherry 1992). This lack of model application to actual field situations is likely due to the fact that field data sets are rarely sufficient for representation using DFMD models. The data sets typically acquired from field sites do not warrant use of such powerful models; simpler analytics can be adequate for examination of solute transport scenarios involving single or parallel plate fractures. Saborin (1989) and Burke (1997) provide examples of use of parallel-plate solute transport models to assess downward contaminant and atmospheric tritium migration, respectively, in clayey aquitards in Canada.

Jorgensen and colleagues in Denmark have made intensive use of FRAC3DVS to represent contaminant migration in surficial clay-till aquitards in Denmark. The Danish researchers have conducted solute tracer experiments in large cylindrical columns of clay till and also at field sites. Klint (2001) mapped fractures to depths between 3 – 8 m below ground surface in large excavations at thirteen sites distributed across Denmark. All of the fracture mapping occurred in basal tills deposited underneath progressing glaciers and consequently had a high potential for development of tectonic fractures. Jorgensen, Klint, and Kistrup (2003) used FRAC3DVS and information about fracture occurrence from these observations to assess the probability of conventional vertical monitoring wells intersecting fractures in aquitards with representative fracture distributions and properties. They evaluated pesticide-monitoring results from different positions of monitoring well screens relative to fractures. They showed that underlying aquifers can be subjected to contamination by downward moving contaminations without being observed in monitoring wells in the till. Jorgensen et al. (2002) used FRAC3DVS to represent vertical flow and pesticide transport along fractures in water-saturated unoxidized clayey till based on tracer experiments and pesticide distribution data from two experimental agricultural fields on clay till underlain by a regional limestone aquifer. Using effective fracture spacings and mean fracture apertures derived from field observations, they found that the model provided reasonable approximations of the concentrations observed in the aquifer wells.

Jorgensen et al. (2004b) used FRACTRAN to represent nitrate transport and denitrification observed in large column laboratory experiments using fractured clay till. Jorgensen et al. (2004a) used a dual porosity EPM model and FRACTRAN to represent laboratory bromide tracer results from four large clay till columns and found that only FRACTRAN was capable of representing the experimental observations without need to adjust fracture spacing and aperture to force model fits. Jorgensen, McKay, and Kistrup (2004) used FRAC3DVS to assess the influence of key factors, mainly recharge and degradation half life, on the downward movement of a widely used pesticide (mecopxp) through a typical Danish clay-till aquitard. The model was calibrated with laboratory and field data from a particular site in Denmark but the overall findings are expected to be relevant to many other clay-till sites in similar settings.

Effective Porosity Values for Calculating Average Linear Groundwater Velocity

Calculation of the average linear velocity to obtain time of water travel from source to receptor using Darcy's Law requires knowledge of the effective porosity (Equation 3.4). Conversion of the Darcy flux to average linear groundwater velocity ($\bar{v}$) in granular deposits (i.e., sand or gravel aquifers) is straightforward because many laboratory and field tracer experiments confirm total porosity provides accurate velocity values and because total porosity is easily estimated or measured. However, hydrogeologists debate the nature and magnitude of the effective porosity appropriate for the calculation of velocity in unfractured and fractured aquitards and, more generally, in all fractured geologic media. For sandy aquifers, the uncertainty in the effective porosity is small, generally small (e.g., the porosity range is generally 0.3 ±0.1), but in fractured media the uncertainty can be orders of magnitude. Thus, in fractured media where $\bar{v}$ is much larger than the Darcy flux, the uncertainty in groundwater travel times can be correspondingly large.

Different effective porosity values appear appropriate for different situations, depending on several factors including the nature of the aquitard (unlithified clay or bedrock), the contaminant, and the nature of the fracture network. In some cases, the effective porosity is used as a "model fitting parameter" to account for the effects of preferential pathways and matrix diffusion. As such, the effective porosity is not an actual physical or measurable porosity. The approach that one takes to obtain $\bar{v}$ and the associated uncertainties are particularly important in investigations of potential contaminant transport and wellhead protection in fractured bedrock using numerical particle tracking (Rayne, Bradbury, and Muldoon 2001).

In fractured aquitards, the calculated $\bar{v}$ values for individual fractures or fracture networks have most direct relevance to the migration of particulate contaminants not retarded by matrix diffusion. Particulate contaminants can be filtered in the fracture network causing immobilization of many or all particles, however, those particles not filtered out can be expected to migrate more or less at the average rate of groundwater flow. McKay et al. (1993) demonstrated very rapid horizontal transport of colloid tracers in a field experiment in a the weathered, intensely fractured zone of a surfical clayey aquitard. The colloid-size tracers (bacteriophage) moved much faster than the solute tracers (bromide and chloride) because the solutes were strongly retarded by matrix diffusion nd the colloid tracers were not. The rapid colloid movement was consistent with the high $\bar{v}$ calculated by dividing the Darcy flux by the fracture porosity obtained from the Cubic Law. The mean hydraulic fracture apertures calculated from the Cubic Law were much larger than the diameters of the particles, however apertures are variable along their planes and some fractures have dead ends. Although some of the particles arrived quickly, most did not arrive at all, probably because they were immobilized due to sorption and constrictions in the passageways in the fracture network.

Two points of view exist for the conversion of the Darcy flux to the average linear groundwater velocity in unfractured aquitards. In the first view, the effective porosity is equal to the total porosity measured on core samples by the conventional method (e.g., the gravimetric method), and consequently all of the water present in this void space (i.e., the total porosity) is considered mobile when a hydraulic gradient is applied. Alternatively, in the second view, much of the water is immobile except for the water in the largest connected pores; therefore, the calculated $\bar{v}$ from the Darcy flux is much larger because the same discharge must flow through a much smaller cross-sectional area. Laboratory column experiments using non-reactive dissolved tracers can help resolve the disagreement, however, to our knowledge the literature reports only one study (Desaulniers 1986) using aquitard material. In this study, conducted using cylindrical columns, two tracers (^{3}H and ^{36}Cl) were passes through cores from two water-saturated clayey aquitards. In the case of silty glaciolacustrine clay, the effective porosity calculated from the $C/C_o = 0.5$ point on the tracer breakthrough curve provided essentially the same value (~0.46) as the total porosity obtained gravimetrically. However, a sample of glaciomarine clay showed slightly faster chloride (^{36}Cl) tracer breakthrough attributed to anion exclusion effects. Anion exclusion is the process in which the electrical charges on the clay particles and on the ions act to prevent some ions from moving as fast as the uncharged water molecules. To match the ^{36}Cl tracer breakthrough, a somewhat lower effective porosity (0.48) was needed compared to the porosity value (0.64) determined gravimetrically. In practical terms, the difference between the total porosity and the value obtained from the breakthrough graph is not important relative to the other uncertainties typically inherent in velocity estimates. These laboratory tracer-breakthrough tests are simple in concept but are extremely time-consuming to conduct because the rate of flow through unfractured natural clay or shale is so slow, typically on the order of a few millimeters per day or less under normal hydraulic gradient.

Another line of evidence provides insight for assessing the two views on effective porosity for $\bar{v}$ calculation in aquitards. Vertical profiles of natural tracers such as ^{18}O, ^{2}H or chloride have been obtained from clayey aquitards in several regions by sampling piezometers or squeezing water from core samples (e.g., Birks 2004, Desaulniers et al. 1986, Simpkins and Bradbury 1992, Husain et al. 1998)). At locations where the clayey aquitards are thick, and composed of glacial till or glaciolacustrine deposits, these stable isotope profiles commonly show gradual differences with depth, from shallow values typical of present-day precipitation to deep values representing the much colder climatic conditions that occurred in late Pleistocene time. These profiles are compared to simulated profiles using mathematical models that represent groundwater flow (advection) and diffusion over time scales of many thousands of years. For clayey aquitards that show no evidence of fractures below the weathered zone, the simulated profiles are controlled by diffusion with small or negligible advection, and commonly match the measured profiles most closely when $\bar{v}$ is obtained through calculations using the total porosity rather than a smaller "effective porosity". In most cases, the best matches are obtained when the $\bar{v}$ value assigned to the model is insignificant (e.g., less than a few millimeters per year). Values of this magnitude are derived from Equation 3.4 for field sites using measured values of total porosity, hydraulic conductivity from lab tests, and field values for hydraulic gradients in the aquitard. When much smaller porosity values are used, the calculated $\bar{v}$ values typically are too large to produce a close match between the field and simulated ^{18}O or ^{2}H profiles and therefore these smaller porosity values are probably not realistic.

The debate regarding approaches for obtaining groundwater velocity values in fractured aquitards has different relevance, depending on the type of contaminant under consideration. Groundwater velocity influences on migration of dissolved contaminants (i.e., solutes) are much different than on particulate contaminants such as bacteria and colloids that may travel at rates represented by $\bar{v}$. The solute migration rate is commonly retarded relative to $\bar{v}$ by the matrix diffusion effect. For aquitards in which solute behavior is diffusion-controlled, the calculated $\bar{v}$ value is relevant only to the degree necessary to establish the aquitard is diffusion-controlled. At some sites, the evidence indicating diffusion-control is derived from isotope and/or major ion data and the evidence generally overrides whatever $\bar{v}$ values result from the Darcy-based calculations. This is the case when the uncertainties in the results of the solute evaluation are smaller than those in the $\bar{v}$ calculation.

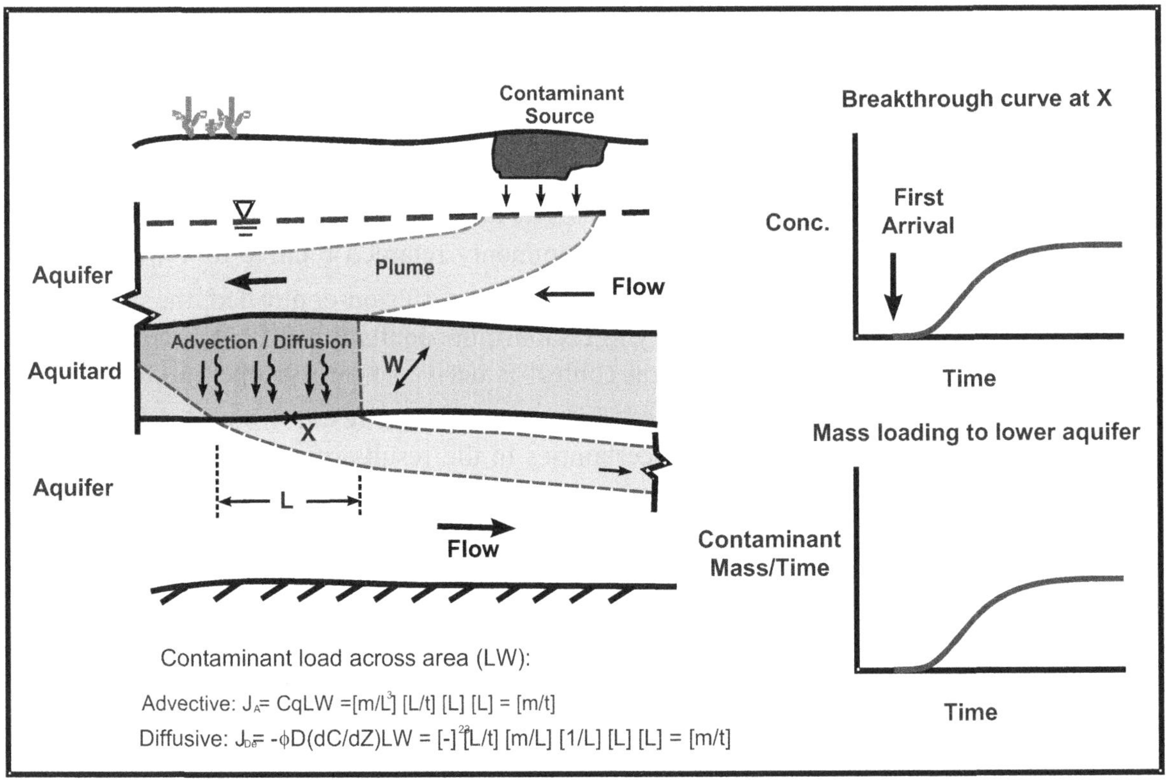

Figure 4.1 Illustration of contaminant pathway from the surface through an aquitard: contaminant breakthrough curve for a point at the top of the lower aquifer and contaminant loading curve for the contribution of contaminants from the aquitard to the aquifer

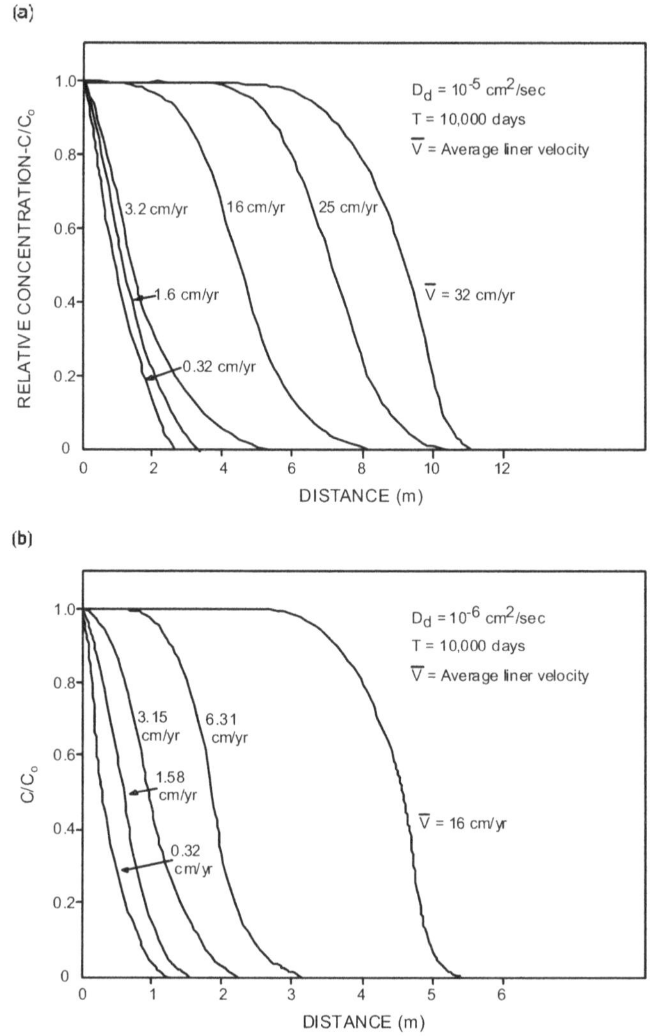

Source: Gillham and Cherry 1982

Figure 4.2 Advance of a solute front along a flow path as a result of advection and molecular diffusion. a) Effective diffusion coefficient Dd = $1*10^{-5}$ cm²/s. b) Effective diffusion coefficient Dd = $1*10^{-6}$ cm²/s. Larger diffusion coefficient causes diffusion to dominate over advection at large gw velocities

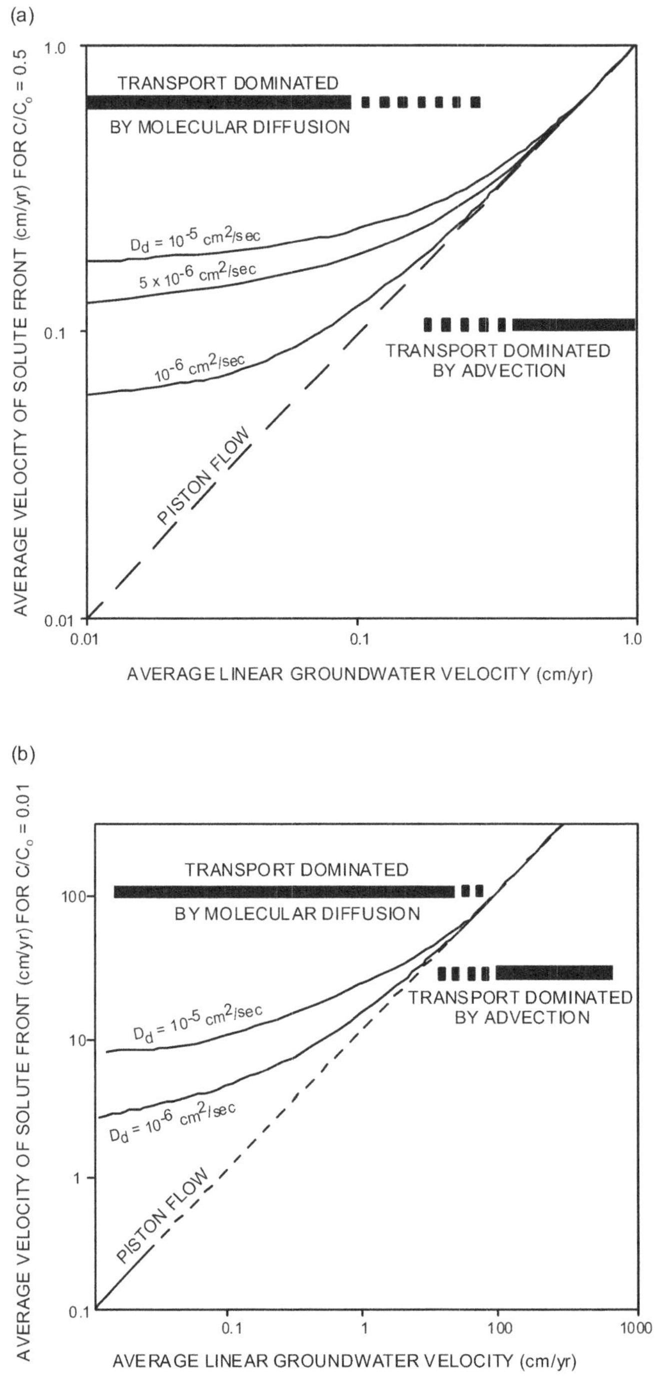

Source: Gillham and Cherry 1982

Figure 4.3 Comparison of the average linear groundwater velocity and rate of solute front advance along a flow path with advection and molecular diffusion. a) The front is defined by C/C0 (i.e., the middle of the front). b) The front is defined by C/C0 = 0.01 (i.e., near the leading edge of the front). Defining the front at a lower concentration causes diffusion to be dominant over a larger velocity range

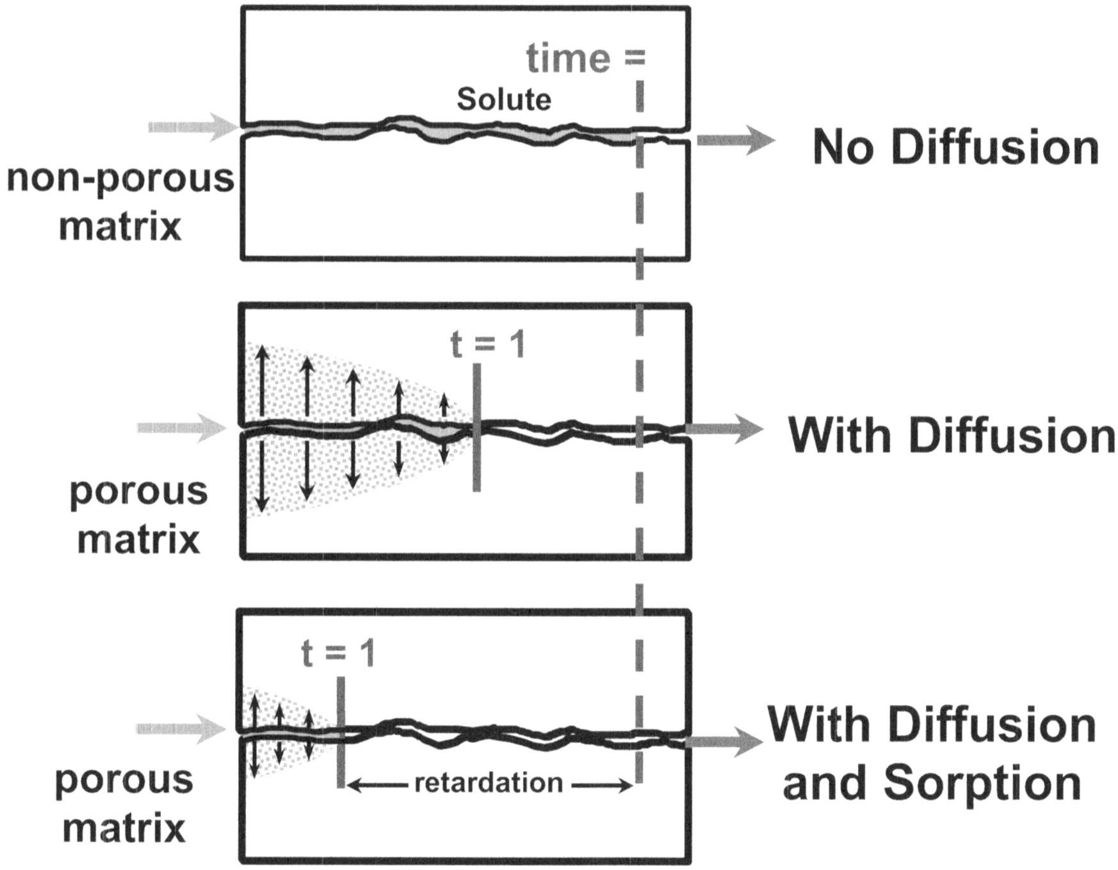

Source: Modified from Freeze and Cherry 1979. *FREEZE, ALAN R.; CHERRY, JOHN A., GROUNDWATER, 1st ©1979.* Adapted by permission of Pearson Education, Inc., Upper Saddle River, New Jersey.

Figure 4.4 Illustration of the concept for solute retardation in fracture caused by molecular diffusion: a) solute transport in a single fracture in a geologic medium with no matrix porosity (e.g., granite), b) solute transport in a porous medium allowing diffusion-driven chemical mass transfer of solute mass from the fracture into the matrix; this causes retardation of the advance of the solute front relative to the rate of plus flow of water, and c) same as case b), but sorption occurs in the porous matrix, which causes more rapid diffusion-driven mass transfer and greater retardation of the solute front.

CHAPTER 5

BIOLOGICALLY ACTIVE CONTAMINANTS IN AQUITARDS

INTRODUCTION

Among the variety of human pathogenic microorganisms transmitted via water, enteric viruses have the greatest potential to move deeply through the subsurface environment, penetrate an aquitard, and reach a confined aquifer. Enteric viruses are extremely small (27 – 75 nm; 1000 nm = 1 μm), allowing them passage through sediment pores that would trap much larger bacteria and protozoa. Depending on sediment chemistry, they may not be completely attenuated by adsorbtion onto sediment grains. Viruses have been detected in groundwater at depths of 67 m (220 ft) (Keswick and Gerba 1980) and 52 m (171 ft) (Borchardt et al. 2003), and reported to move laterally as far as 408 m (1339 ft) in glacial till and 1600 m (5300 ft) in fractured limestone (Keswick and Gerba 1980, Robertson and Edberg 1997). Their transport potential is further evidenced by their widespread occurrence in drinking water wells (Gerba and Rose 1990). In one study of 448 groundwater sites in 35 states, 141 sites (31.5%) were positive for at least one virus type (Abbaszadegan, Lechevallier, and Gerba 2003). Enteric virus types found in groundwater include enteroviruses, hepatitis A virus, rotavirus, and noroviruses (Abbaszadegan, Lechevallier, and Gerba 2003; Borchardt et al. 2003; Fout et al. 2003). Astroviruses and adenoviruses 40/41 are relatively newly discovered enteric viruses documented to occur in surface water (Chapron et al. 2000, Van Heerden et al. 2003) and will likely be found someday in groundwater. From a public health perspective, more than half of waterborne disease outbreaks in the USA are attributable to drinking groundwater, about half of which have a viral etiology (Craun, Berger, and Calderon 1997; Lee et al. 2002; U.S. EPA 2000). For these reasons, this section focuses on viruses as microbial contaminants through aquitards.

VIRUS OCCURRENCE IN CONFINED AQUIFERS

We are not aware of any published studies whose primary objective was to sample confined aquifers for enteric viruses. The only study with some information is Powell et al. (2003) who used multilevel piezometers to take depth-specific samples from five deep sandstone aquifers, one of which was confined by fine-grained deposits of the Mercia Mudstone Group. In this aquifer, samples from a depth of 91 m were positive for coliphage virus, coliform bacteria, fecal streptococci, and clostridia spores; none of the samples at this depth were positive for enteroviruses. From another aquifer, apparently classified as unconfined by the authors even though the borehole log showed several mudstone bands, samples from 47 m were positive for coliforms, fecal streptococci, coliphages, and culturable enteroviruses. The piezometers were located in urban areas and the authors suggest the contamination resulted from leaking sewers and microscopic groundwater flow along fissures and bedding planes. The depths of microbially contaminated samples appeared to correspond to the depths of fissures observed in the boreholes.

The Proposed Groundwater Rule (U.S. EPA 2000) summarizes several studies on the occurrence of pathogens and indicator micro-organisms in groundwater that mention sampling from confined aquifers. Unfortunately, no data are reported and the studies are unpublished or were incomplete at the time of the proposed rule. In the study by (Abbaszadegan, Lechevallier, and Gerba 2003) on the prevalence of enteric viruses in US groundwaters, approximately 25% of the wells sampled were deeper than 500 ft and presumably some of these wells withdrew water from confined aquifers. The American Water Works Research Foundation funded this study, the well logs for the sampled wells are reportedly available (U.S. EPA 2000), and thus it may be possible to revisiting the data and determine if any of the virus-positive samples were from confined aquifers may be possible.

CLAYEY AQUITARDS

Clayey aquitards, while generally having low hydraulic conductivity, can contain fractures and root holes large enough to allow passage of viruses into underlying aquifers. Fractures and root holes can be as large as 100 μm, 5000 times larger than the smallest viruses (Hinsby et al. 1996). Viruses pass through fractured clay till in apertures as small as 3 to 5 μm (Sims 1993). When a colloid, like a virus, is introduced to clay, the colloid preferentially flows through the large apertures offered by fractures and root holes because it essentially cannot enter the much smaller pores between clay grains. In one study, about 50% of clay pore throats were less than 65 nm, about the same size as enteric viruses (McKay, Gillham, and Cherry 1993). Preferential flow results in colloids moving substantially faster than dissolved solutes like bromide because the solute will enter the smaller porous region with the more tortuous, longer, and ultimately slower flow paths. McKay, Gillham, and Cherry (1993) demonstrated that the viruses PRD-1 and MS-2 move through fractured clay 100 to 200 times faster than bromide and the virus velocities were 2 – 5 m/day. Virus levels were attenuated, perhaps due to filtering in fracture constrictions or adsorption to the fracture walls. Hinsby et al. (1996) introduced PRD-1 and MS-2 to clay with fractures and root holes and measured the virus velocities as 4 – 360 m/day (13-1180 ft/d). Fractured aquitards may be more vulnerable to virus contamination than unfractured aquitards not as extensively traversed with fractures and root holes, although this is conjecture since we are not aware of virus transport studies using other aquitard types.

VIRUS TRANSPORT MODELS

Although the literature on groundwater transport of viruses is extensive, the state-of-the-science is not sufficient to adequately predict whether an aquifer or particular well will become virus-contaminated. Mathematical models representing virus transport are available publicly, such as VIRTUS (Yates and Ouyang 1992), CANVAS (Park et al. 1994), and HYDRUS_2D (Simunek, Sejna, and van Genuchten 1999). These models are based on the primary processes of virus transport: advection, dispersion, sorption, and inactivation (i.e., virus decay). However, these processes are highly heterogeneous on a field-scale and differ widely by virus type resulting in model outputs with a great deal of uncertainty (Yates and Jury 1995). Yates et al. (2000) compared predicted virus concentrations from CANVAS and HYDRUS_2D in a septic system leach field. CANVAS predictions either overestimated or underestimated the field data. HYDRUS_2D accurately predicted the virus breakthrough curves, however, the model requires extensive input data and advanced expertise not generally available and very expensive to obtain. According to Yates, virus transport models are either too complex or the model predictions too uncertain for routine use by drinking water utilities for determining whether wells are in regulatory compliance for acceptable virus concentrations; the models should be reserved for research purposes only (personal communication, M.V. Yates, 2004). Recently, the U.S. Environmental Protection Agency developed a user-friendly, Monte-Carlo simulation based screening model called VIRULO, where the user specifies an acceptable virus groundwater concentration and the model outputs the probability of the unsaturated zone attenuating the viruses to the target level (Faulkner et al. 2002). We are not aware of any field validation studies of this model, although perhaps VIRULO could be used to determine the probability that the saturated zone above an aquitard will become virus contaminated. A separate model would then be needed for predicting virus movement through the aquitard into the confined aquifer, but to our knowledge this type of model does not exist. Aquitards are rarely part of the conceptual framework for virus subsurface transport (cf. McKay, Gillham, and Cherry 1993).

FACTORS INFLUENCING VIRUS TRANSPORT AND SURVIVAL

In the face of the uncertainties in the virus transport models and their limited applicability to aquitards, how can the vulnerability of confined aquifers to contamination from human enteric viruses best be assessed? Some kind of decision tree or vulnerability score might be options, but the usefulness of even these simple models would still need to be validated, and at this time the empirical data on virus occurrence in confined aquifers is nearly non-existent. Instead, the approach in this report is to summarize those factors known to be major determinants of virus transport and for which, for the most part, site-specific information is available to drinking water managers. With this information in hand, managers can use their experience and knowledge of their groundwater system to develop an informed opinion on the likelihood for viruses to reach the confined aquifer and whether virus testing would be prudent. If the vulnerability of a confined aquifer to virus contamination is deemed to be low, contaminatation with lower-transport-potential pathogens like bacteria and protozoa is also unlikely (Table 5.1). More comprehensive and detailed information on virus transport and survival can be found in reviews by Gerba and Bitton (1984), Yates and Yates (1988), Bitton and Harvey (1992), Schijven and Hassanizadeh (2000), and Azadpour-Keeley, Faulkner, and Chen(2003).

Contamination Source

In order for a well to become virus-contaminated, there must be a fecal source nearby, as enteric viruses can only multiply within a host and are released from their host via fecal excretion. Potential fecal sources include leaking sanitary sewer lines, septic systems, landfills, field-applied sludge or septage, effluent holding ponds, wastewater irrigation sites, injection wells, reclaimed water application or recharge sites, and surface waters infiltrating the well. Animal manure can also be a source of bacterial pathogens, for example *E. coli* O157:H7 and some human enteric viruses, for example, rotavirus. Clearly, the greater the fecal loading onto or into the ground, the greater the opportunity for the underlying groundwater to become virus-contaminated. A well located in the middle of a large city underlain with leaking sanitary sewers stands a greater chance of becoming pathogen-contaminated than a well adjacent to a few septic systems. Intuitively, the greater the level of fecal loading near the ground surface, the greater the level of viruses reaching the upper boundary of an aquitard, and the greater the opportunity for viral movement through the aquitard. Preventing virus contamination of a confined aquifer begins with decreasing fecal loading above the aquifer and increasing the distance from a fecal source to a drinking water well. What constitutes an effective distance between a fecal source and well drawing from an unconfined aquifer is largely educated guesswork, given the uncertainties in predicting viral transport. Add an aquitard element, for which there is even less virus transport information, and the appropriate setback distance between fecal source and a well drawing from a confined aquifer is more uncertain.

Table 5.1 Characteristics of biological contaminants in groundwater

Biological contaminant[1]	Typical size	Metabolic state in groundwater	Frequency in unconfined aquifers	Survival time in groundwater
Viruses	27 – 75 nm	Infectious, but cannot replicate without host	5 – 30% of wells in USA	1 – 2 years
Bacteria	0.5 – 2 μm	Infectious, potentially replicating in water	Less frequent than viruses	Months
Protozoa	4 – 30 μm	Infectious, environmentally resistant cyst, cannot replicate without host	Rare, unless surface water influence	Unknown

[1]Refers to human pathogenic forms

Water Table Depth

The unsaturated zone is generally more effective in removing viruses than the saturated zone (Chu and Jin 2001; Hurst et al. 1980; Lance and Gerba 1984b; Poletika, Jury, and Yates 1995; Powelson and Gerba 1994). Therefore, as the depth of the unsaturated zone increases (i.e., depth to water table increases), more and more viruses will be removed and fewer will be available to potentially breach an aquitard. Fecal contamination sources that release viruses directly into the saturated zone, such as injection wells, surface water infiltration, or sewer lines placed at a depth where the water table seasonally inundates, are the least attenuated because the viruses are never exposed to an unsaturated zone.

Pore Water pH

Subsurface transport of viruses is attenuated by adsorption of virus particles to sediment grains. The mechanisms responsible for the adsorption process are van der Waals forces, hydrogen bonding, hydrophobic interactions, and electrostatic interactions, of which the last mechanism appears the most important (Dowd et al. 1998; Guan et al. 2003; Redman, Grant, and Olson 1997). Viral capsids (the protein shell) are charged on the outer surface, and when the virus collides with a sediment grain with an opposite charge, the virus is adsorped. Water pH determines the charge on the sediment grain and virus. At pHs less than 7.0, viruses tend to be more positively charged and will sorp strongly to sediment grains which tend to be negatively charged. The exact charge on a virus at a particular water pH depends on the virus type, but generally viruses will travel less distance in acidic water compared to water of neutral or alkaline pH(Goyal and Gerba 1979).

Sediment Texture

The continuum of sediment textures from coarse to fine is gravel, sand, silt, and clay and of these, viruses move most readily in gravel and sand (Gerba and Bitton 1984, Goyal and Gerba 1979, Yates and Yates 1988). Sediments containing gravel and sand have large pore throats for easy passage of virus particles, and moreover, viruses experience fewer collisions with the large grains meaning fewer opportunities for adsorption. Sand can electrostatically adsorp viruses, particularly sand containing metals and metal oxides (Chu and Jim 2001). Cations with high valencies, for example Mg^{+2} and Ca^{+2}, are believed to increase virus adsorption (Redman et al. 1999, Yates and Yates 1988). Mineral colloids, such as the clay minerals montmorillonite and kaolinite enhance virus transport through sand (Jin, Pratt, and Yates 2000). However, fewer viruses will be removed during passage through sand than through clay. Clay can physically filter viruses because of its small pore throats, presentation of maximum grain-virus collisions and opportunities for adsorption, and because of high charge, viruses are strongly adsorped by electrostatic forces. Therefore, confined aquifers overlain with coarse textured sediments may be more susceptible to virus intrusions than those overlain with silt and clay.

Ionic Strength and Rainfall

The ionic strength of the pore water solution is a major factor in the adsorption and desorption of viruses to sediment grains (Chu, Jin and Yates 2000, Lance and Gerba 1984a). For the purposes of this report, ionic strength is considered the concentration of anions and cations in solution (for a complete definition see Snoeyink and Jenkins 1980). Generally, as ionic strength increases so does the strength of adsorption. Adsorption is not necessarily irreversible and desorption does occur. One event associated with virus desorption and enhanced transport is a decrease in ionic strength (Bales et al. 1993; Duboise, Moore, and Sagik 1976). Rainfall may enhance virus transport because of its low ionic strength (Bitton and Harvey 1992, Gerba and Bitton 1984); confined aquifers in regions with heavy rainfall may be more likely to be virus-contaminated compared to those in arid regions.

Organic Matter and Surfactants

Generally, when dissolved organic matter (DOM) concentration is elevated in sediments, virus transport is enhanced (Pieper et al. 1997; Powelson, Simpson, and Gerba 1991), probably because DOM blocks electrostatic interactions between viruses and sediment grains. However, a few studies report organic matter will retard viral transport (Bales et al. 1993); a suggested explanation is that DOM increases the hydrophobicity of sediment grains and hydrophobic interactions are another mechanism of virus adsorption (Kinoshita et al. 1993). These study findings may be discordant because of unknown differences in experimental design, such as virus type, DOM type, or sediment characteristics. DOM is routinely used in the laboratory to elute viruses from electropositively charged filters and highly organic muck soils do not adsorp viruses well (Bitton and Harvey 1992), suggesting a positive association between DOM concentration and virus transport is probably more the rule than the exception. Surfactants common in household products and therefore in wastewater appear to have two opposite effects on viruses: 1) diminishing virus survival, and 2) enhancing virus mobility (Chattopadhyay et al. 2002). The overall effect of surfactants on virus contamination of groundwater is not clear at this time. In general, fecal contamination sources with high levels of DOM and some level of surfactants, such as septic systems and leaking sanitary sewers, present a greater potential for virus transport than sources with lower DOM concentrations, such as reclaimed water for irrigation.

Virus Survival

For people to become virally infected from drinking contaminated groundwater, not only do the viruses need to be transported to the well, they must also survive the transport process and remain infectious. The most important determinant of virus inactivation in groundwater is temperature; as temperature increases so does the inactivation rate (Hurst, Gerba, and Cech. 1980; Yates, Gerba, and Kelley 1985; Yates and Yates 1988). Yates, Gerba, and Kelley (1985) showed MS-2 virus incubated in Arizona groundwater is inactivated at nearly 10 times the rate at 23°C compared to 4°C. This same study estimated at groundwater temperatures found in most areas of the United States, which are typically low, nearly a month is required to achieve a one log reduction in the concentration of poliovirus 1 and echovirus 1. Inactivation rates differ by virus type; hepatitis A virus is one of the more persistent enteric viruses at high temperatures (Blanc and Nasser 1996, Sobsey et al. 1986). Soil moisture is also an important determinant with viruses surviving longer in moist soils (Yates and Yates 1988). Therefore, confined aquifers in arid regions with warm groundwater temperatures (e.g., 20°C) may be less prone to contamination with infectious viruses.

Another consideration is the travel time for a virus to reach a confined aquifer. If the travel time is greater than the virus survival time, the virus will unlikely be infectious when it reaches the well. The upper limit for survival time in groundwater has not been determined for many viruses. Estimates can be made from inactivation rates and assuming an initial virus concentration, but have rarely been observed empirically. Poliovirus survived for more than six months in saturated sand at 4°C (Yeager and O'Brien 1979) and in river water at 4°C (Rhodes et al. 1950); it was also observed to be inactivated by only 99% (i.e., two log reduction) after incubation for 175 days in sand at 1 to 8°C (Lefler and Kott 1974). Given the inactivation rate of hepatitis A virus is less than poliovirus, a reasonable estimate of the longest survival time for enteric viruses as a group would be one to two years. Water taking 10 to 100 years to reach a confined aquifer is unlikely to transport infectious viruses. However, estimating groundwater travel time accurately is difficult, and measurements of groundwater age of the bulk fluid may not reflect microscopic contributions via preferential flow that, albeit small, contain measurable infectious viruses (Powell et al. 2003).

Other Factors

Other important factors involved with virus transport and survival are summarized briefly here. Site-specific information for these may be less readily available to drinking water managers than for previously presented factors, rendering them less relevant for assessing confined aquifer vulnerability unless site-specific studies are undertaken.

Virus type will determine transport potential as each differs in important transport characteristics such as size and isoelectric point (pI). The pI is the water pH that balances the charges on the virus capsid to a net charge of zero. At a pH above its pI, the virus has a net negative charge; at lower pH, the net charge is positive. Dowd et al. (1998) measured the transport of five viruses with pIs ranging from 3.9 to 7.3 in sandy soil with a pH of 7.1 and found virus pI and adsorption level were inversely related. pI was the most important virus-specific characteristic determining transport, although size was important for viruses with diameters larger than 60 nm. Similarly, Woessner et al. (2001) seeded four virus types into an alluvial sand-gravel aquifer at a location 21.5 m from a pumping well and found the proportions of seeded viruses recovered at the well were correlated with the virus isoelectric points.

Antagonism from indigenous bacteria and protozoa appears to be one mechanism of virus inactivation in groundwater. Gordon and Toze (2003) compared survival times of poliovirus and coxsackievirus seeded into filtered (0.2 μm) and unfiltered groundwater. In the absence of indigenous micro-organisms the viruses survived 30 to 40 times longer than in unfiltered ambient water. This microbial effect was stronger than temperature in determining virus decay rates, and the investigators suggest the generally accepted positive correlation between temperature and virus inactivation may be due, in part, to higher temperatures enhancing microbial metabolic activity.

This brief review has not discussed potential interactions among factors, which are probably important but not fully understood.

CONCLUSIONS AND RECOMMENDATIONS

At this time, predicting whether a particular aquitard will prevent adjacent aquifers from becoming virus contaminated is not possible. A virus transport model may be useful for approximating the virus load moving through the unconfined aquifer and reaching the aquitard upper boundary. Alternatively, drinking water managers might do just as well in judging the likelihood of viruses reaching the aquitard based on their knowledge of the virus transport and survival factors pertaining to their local groundwater systems. A worst case system would be one with a fecal source with high loading rates near the drinking water wells, a shallow water table in sandy soil, sediment pore water with a neutral to alkaline pH and a high dissolved organic matter level, plenty of low ionic strength precipitation, and cool groundwater temperatures. However, even in this case where virus transport through an unconfined aquifer would appear optimal, the quantity of viruses penetrating the aquitard is unknown. Some knowledge of the aquitard's characteristics (e.g., extent of fracturing) and / or time of travel from source to receptor (months versus few years to decades) is necessary to arrive at an informed opinion on the vulnerability of the confined aquifer. Even then, as the National Research Council has written, "… since all [groundwater] vulnerability assessments are uncertain, no management decisions based on them are ever clear-cut or certain." (National Research Council 1993). The most confident assessment and assurance of sanitary quality would come from sampling a confined aquifer for pathogenic enteric viruses, preferably monthly through all four seasons. This approach will be more feasible in the future as methods for testing waterborne viruses become more standardized, less complicated, and more affordable. In the meantime, sampling a confined aquifer for coliphages (i.e., viruses of coliform bacteria) would be somewhat informative because they are shed in human feces, tests for their detection are inexpensive, and their presence in a confined aquifer suggests pathogenic viruses could also be breaching the aquitard. However, the absence of coliphages does not necessarily indicate absence of pathogenic viruses in the same aquifer. Ultimately, more data are required on virus occurrence in confined aquifers, and more studies are needed to define the relationship between virus presence and specific, readily measured aquitard characteristics.

CHAPTER 6

SUMMARY AND CONCLUSIONS

SUMMARY

This document contains a comprehensive review of the state of the science of aquitard studies, with emphasis on factors and issues concerning the protection that aquitards provide for adjacent aquifers and wells used for water supply. This report considers many aspects of aquitard science, including geologic settings, groundwater flow, contaminant transport, virus transport, heterogeneities, and predictive modeling.

CONCLUSIONS

Preparation of this report has led the authors to a number of conclusions, as follows:

Current Understanding of Aquitards

- The traditional hydrogeologic and engineering literature usually presents an oversimplified treatment of aquitards. Textbooks depict aquitards as uniform units of considerable horizontal extent, but most field studies demonstrate aquitards typically have large spatial variability in thickness and lithology. They also commonly contain secondary features such as fractures, macropores, or other heterogeneities that may strongly influence contaminant migration.

- Shallow unlithified aquitards have received significantly more scientific attention than deep bedrock aquitards, especially in studies related to contaminant migration. This bias is understandable because shallow unlithified aquitards are more immediately related to shallow groundwater contamination issues, are relatively accessible for study, and are less expensive to instrument than deeper bedrock aquitards. There is a need for additional studies of deep bedrock aquitards, especially where these aquitards occur above major water-supply aquifers.

- Hydrogeologists should designate geologic units or zones as aquitards based on hydraulic data rather than stratigraphy or lithology. Aquitards are represented in textbooks and reports as being composed of entire geologic formations or stratigraphic units when this is rarely the case. Often only a few key beds within a much thicker stratigraphic unit provide the restrictions in flow and/or contaminant migration to warrant designation as an 'aquitard'. As a consequence, aquitards can be thinner, and adjacent aquifers thicker, than stratigraphic considerations alone would indicate.

- The most useful information about an aquitard is usually internal data, for example hydraulic head measurements or water samples collected from piezometers situated inside the aquitard. External measurements, such as hydraulic head measured in aquifers above and below an aquitard, provide only indirect evidence of aquitard performance. Many commonly-used methods for characterizing the hydraulics of aquitards (for example, leaky-aquifer pumping test analyses) are of limited value in predicting contaminant movement through aquitards, because the inherent simplifying assumptions are often inappropriately applied in the analyses of contaminant movement. These external methods rarely identify the types or location of migration pathways in aquitards.

- The bulk hydraulic properties of aquitards are generally very poorly known. Direct field measurements, at appropriate scales, of important hydraulic parameters such as vertical hydraulic conductivity and effective porosity are usually necessary to assess aquitard effectiveness and predict contaminant movement.

- Of the various types of contaminants that threaten aquifers, DNAPLs have the greatest propensity to move through fractured aquitards because i) their movement in fractures is not likely to be retarded by matrix diffusion and sorption, and ii) they can flow against the hydraulic gradient, except when the counterflow hydraulic gradient is large. Particulate contaminants very small in size, such as viruses, have the next largest propensity to travel quickly through fractured aquitards because they are not retarded by matrix diffusion. However, their downward movement is prevented by an upward hydraulic gradient. Also, viruses may be attenuated by filtering and sorption. Dissolved contaminants have the least propensity to travel through fractured aquitards because their migration is retarded by matrix diffusion and other attenuating processes, such as sorption, precipitation and / or degradation.

- The literature identifies various types of preferential pathways for contaminant migration through aquitards, however only minimal guidance is provided on how to go about conducting field studies to locate and characterize these pathways.

Vulnerability of Water Supply Wells

- Aquitards are critical to protecting water supply wells from contamination. In general any well constructed with no aquitard between a contamination source and the well screen (or open borehole) is at great risk.

- Hydrogeologic and engineering studies conducted when designing water supply wells should strive to collect sufficient data on adjacent aquitards to allow a reasonable assessment of vulnerability to contamination and prevent well designs that cross connect or breech aquitards.

- With respect to the protection that aquitards offer for water supply wells finished beneath them, assessments of the risk or probability of contamination will be appropriate, meaningful, and useful only if the likely contaminant pathways through the aquitard are identified and characterized.

- The extremely small size of viruses (<100 nm) compared to the probable sizes of fractures and macropores observed in aquitards (5-50 μm) suggests aquitards may not always provide barriers to virus transport and research is needed to address gastrointestinal virus transport and survival through aquitards.

- Whether a particular aquitard will prevent adjacent aquifers from becoming virus contaminated is not always possible to predict. The most confident assessment and assurance of sanitary quality would come from sampling a confined aquifer for pathogenic enteric viruses, preferably monthly through all four seasons.

Opportunities for Improved Aquitard Assessment

- Appropriate hydrogeologic and geophysical tools and methods for assessing aquitards have been developed and are now available for systematic application in standard practice. However, at present only a few aquitards have been investigated using many of these tools. Progress has been hindered by the lack of textbooks and guidance documents focused on aquitards. A guidance document prepared as a companion to this report is intended to fill part of this literature deficiency.

- Current aquitard studies are heavily biased toward shallow, unlithified aquitards. Although scientific investigation of shallow unlithified aquitards should continue, there is a need for similar focused work and increased emphasis on deeper bedrock aquitards.

- The most important field data needed in assessment of the nature of an aquitard is hydraulic head at different depths inside the aquitard. The head measurements should include temporal data.

- Hydrochemical and isotopic constituents of the groundwater both inside and external to the aquitard are commonly valuable long-term inherent tracers diagnostic of aquitard conditions for solute transport.

- Recently-developed innovative investigative sampling and monitoring tools (e.g., borehole imaging, flowmeter logging (where / if appropriate), depth-discrete multilevel monitoring systems, chemical analysis of core samples) can provide essential information for aquitard assessment at reasonable cost and effective mathematical models (i.e., computer codes) are readily available for data assessment. Some of the tools suitable for indurated aquitards are not suitable for non-indurated aquitards.

REFERENCES

Abbaszadegan, M., M. Lechevallier, and C. Gerba. 2003. Occurrence of viruses in US groundwaters. *Journal American Water Works Association,* 95:107-120.

Adar, E.M., and R. Nativ. 2000. Use of hydrochemistry and isotopes in a mixing-cell model to quantify the relative contribution of multiple-source contaminants to seepage from a fractured chalk aquitard. In *Tracers and modelling in hydrogeology* Edited by A. Dassargues. Wallingford, Oxfordshire: IAHS Press

Alley, W.M., R.W. Healy, J.W. LaBaugh, and T.E. Reilly. 2002. Hydrology - Flow and storage in groundwater systems. *Science,* 296:1985-1990.

Alley, W. M., T.E. Reilly, and O.L. Franke. 1999. Sustainability of ground-water resources: U.S. Geological Survey Circular 1186, Denver, Colorado: U.S. Geological Survey.

American Society of Civil Engineers..1960. *Research Conference on Shear Strength of Cohesive Soils*. University of Colorado, Boulder, CO.: Soil Mechanics and Foundation Division

Anderson, M.P. 1989. Hydrogeologic facies models to delineate large-scale spatial trends in glacial and glaciofluvial sediments. *Geological Society of America Bulletin,*101:501-511.

Antonellini, M., and A. Aydin. 1994. Effect of Faulting on Fluid-Flow in Porous Sandstones - Petrophysical Properties. *Aapg Bulletin-American Association of Petroleum Geologists,* 78:355-377.

Aslan, A., and W.J. Autin. 1996. Depositional and pedogenic influences on the environmental geology of Holocene Mississippi River floodplain deposits near Ferriday, Louisiana. *Engineering Geology,* 45:417-432.

Azadpour-Keeley, A., B.R. Faulkner, and J.-S. Chen. 2003. *Movement and longevity of viruses in the subsurface.* OH EPA/540/S-03/500, Cincinnati, OH:.US Environmental Protection Agency, National Risk Management Research Laboratory

Back, W. 1985. Geochemical significance of aquitards. in. *Hydrogeology of Rocks of Low Permeability*. Edited by E.S. Simpson. Tucson, Arizona: International Association of Hydrogeologists (USA Committee): 358-369

Back, W., J.S. Rosenshein, and P.R. Seaber. 1988. *Hydrogeology*. Boulder, Colo: Geological Society of America,.

Bales, R.C., S.M. Li, K.M. Maguire, M.T. Yahya, and C.P. Gerba. 1993. Ms-2 and Poliovirus Transport in Porous-Media - Hydrophobic Effects and Chemical Perturbations. *Water Resources Research,* 29:957-963.

Barker, R.A., P.W. Bush, and E.T. Baker, Jr. 1994. Geologic history and hydrogeologic setting of the Edwards-Trinity aquifer system, west-central Texas. U.S. Geological Survey Water Resources Investigation Report 94-4039.

Barone, F.S., R.K. Rowe, and R.M. Quigley. 1990. Laboratory Determination of Chloride Diffusion-Coefficient in an Intact Shale. *Canadian Geotechnical Journal,* 27:177-184.

Barone, F.S., E.K. Yanful, R.M. Quigley, and R.K. Rowe. 1989. Effect of Multiple Contaminant Migration on Diffusion and Adsorption of Some Domestic Waste Contaminants in a Natural Clayey Soil. *Canadian Geotechnical Journal,* 26:189-198.

Barrash, W. 1986. Hydrostratigraphy and hydraulic behavior of fractured Brule Formation in Sidney Draw, Cheyenne County, Nebraska (Equivalent-porosity medium, aquifer test). Ph.D. diss., University of Idaho.

Barrash, W., and D.R. Ralston. 1991. Analytical Modeling of a Fracture-Zone in the Brule Formation as an Aquifer Receiving Leakage from Water-Table and Elastic Aquitards. *Journal of Hydrology,* 125:1-24.

Bates, R.L., and J.A. Jackson. 1987. *Glossary of geology*. Falls Church, Va.: American Geological Institute.

Batu, V. 1998. *Aquifer hydraulics : a comprehensive guide to hydrogeologic data analysis*. New York: Wiley.

Beauheim, R.L., and R.M. Roberts. 2002. Hydrology and hydraulic properties of a bedded evaporite formation. *Journal of Hydrology,* 259:66-88.

Belitz, K., and J.D. Bredehoeft. 1990. Role of confining layers in controlling large-scale regional ground-water flow. In *Hydrogeology of low permeability environments*. Edited by P. Neuman Shlomo and I. Neretnieks.. Hanover, Federal Republic of Germany: Verlag Heinz Heise,7-17.

Bethke, C.M., and T.M. Johnson. 2002. Paradox of groundwater age. *Geology,* 30:107-110.

Bierkens, M.F P. 1996. Modeling hydraulic conductivity of a complex confining layer at various spatial scales. *Water Resources Research,* 32:2369-2382.

Birks, S.J. 2004. Water isotope partitioning in aquitards and precipitation on the northern Great Plains. Ph.D. diss. University of Waterloo, Waterloo, Ontario.

Bishop, P.K., R. Jakobsen, E. Gosk, D.N. Lerner, and M.W. Burston. 1993a. Investigation of a Solvent Polluted Industrial Site on a Deep Sandstone Mudstone Sequence in the UK .1. Site Description and Groundwater-Flow. *Journal of Hydrology,* 149:209-229.

Bishop, P.K., D.N. Lerner, R. Jakobsen, E. Gosk, M.W. Burston, and T. Chen. 1993b. Investigation of a Solvent Polluted Industrial Site on a Deep Sandstone Mudstone Sequence in the UK .2. Contaminant Sources, Distributions, Transport and Retardation. *Journal of Hydrology,* 149:231-256.

Bitton, G., and R.W. Harvey. 1992. Transport of pathogens through soils and aquifers. Pages 103-124 In *Environmental Microbiology*. Edited by R. Mitchell. New York: Wiley-Liss, Inc.

Blanc, R., and A. Nasser. 1996. Effect of effluent quality and temperature on the persistence of viruses in soil. *Water Science and Technology,* 33:237-242.

Boisson, J.Y., L. Bertrand, J.F. Heitz, and Y. Moreau-Le Golvan. 2001. In situ and laboratory investigations of fluid flow through an argillaceous formation at different scales of space and time, Tournemire tunnel, southern France. *Hydrogeology Journal,* 9:108-123.

Boldt-Leppin, B.E.J., and M.J. Hendry. 2003. Application of harmonic analysis of water levels to determine vertical hydraulic conductivities in clay-rich aquitards. *Ground Water,* 41:514-522.

Bonin, B. 1998. Deep geological disposal in argillaceous formations: studies at the Tournemire test site. *Journal of Contaminant Hydrology,* 35:315-330.

Borchardt, M.A., P.D. Bertz, S.K. Spencer, and D.A. Battigelli. 2003. Incidence of enteric viruses in groundwater from household wells in Wisconsin. *Applied and Environmental Microbiology,* 69:1172-1180.

Bradbury, K.R., D.S. Desaulniers, D.E. Connell, and R.G. Hennings. 1985. Groundwater movement through clayey till, northwestern Wisconsin, USA. In *Hydrogeology of Rocks of Low Permeability*, Edited by E. S. Simpson. Tuscon, AZ: International Association of Hydrogeologists (USA Committee): 405-416.

Bradbury, K.R., and M.A. Muldoon. 1990. Hydraulic conductivity determinations in unlithified glacial and fluvial materials. In *Ground water and vadose zone monitoring*. Edited by D.M. Nielsen and A.I. Johnson. Philadelphia, PA,: American Society for Testing and Materials:138-151

Bradbury, K.R., S.K. Swanson, J.T. Krohelski, and A.K. Fritz. 1999. *Hydrogeology of Dane County, Wisconsin*. Wisconsin Geological and Natural History Survey Open-file Report 99-04. Madison, WI.

Bradbury, K.R., M.B. Gotkowitz, D.J. Hart, T.T. Eaton, J.A. Cherry, B.L. Parker, and M.A. Borchardt, 2006. *Contaminant transport through aquitards: Technical guidance for aquitard assessment*. AwwaRF technical report.

Bredehoeft, J.D., C.E. Neuzil, and P.C.D. Milly. 1983. Regional flow in the Dakota aquifer: a case study of the role of confining layers. U.S. Geological Survey Water Supply Paper 2237.

Brewster, M.L., A.P. Annan, J.P. Greenhouse, B.H. Kueper, G.R. Olhoeft, J.D. Redman, and K.A. Sander. 1995. Observed Migration of a Controlled DNAPL Release by Geophysical Methods. *Ground Water,* 33(6):977-987.

Brockman, C.S., and J.P. Szabo. 2000. Fractures and their distribution in the tills of Ohio. *Ohio Journal of Scienc,e* 100:39-55.

Broholm, K., A.B. Hansen, P.R. Jorgensen, E. Arvin, and M. Hansen. 1999a. Transport and biodegradation of creosote compounds in a large, intact, fractured clayey till column. *Journal of Contaminant Hydrology,* 39:331-348.

Broholm, K., P.R. Jorgensen, A.B. Hansen, E. Arvin, and M. Hansen. 1999b. Transport of creosote compounds in a large, intact, macroporous clayey till column. *Journal of Contaminant Hydrology,* 39:309-329.

Brown, K.W., J.C. Thomas, and J.W. Green. 1984. Permeability of compacted soils to solvents, mixtures and petroleum products. Tenth Annual Research Symposium. EPA-600/9-007, Fort Mitchell, KY:124-137

Brownell, J.R. 1986. Stratigraphy of unlithified deposits in the central sand plain of Wisconsin. Master's thesis. University of Wisconsin-Madison, Madison, Wisconsin.

Burke, M.B. 1997. Depth of active groundwater flow in the fractured clay of the St. Clair Plain, Southwestern Ontario. Master's thesis. University of Waterloo, Waterloo, Ontario.

Butler, J.J., Jr., G.C. Bohling, Z. Hyder, and C.D. McElwee. 1994. The use of slug tests to describe vertical variations in hydraulic conductivity. *Journal of Hydrolog,y* 156:137-162.

Caine, J.S., J.P. Evans, and C.B. Forster. 1996. Fault zone architecture and permeability structure. *Geology,* 24:1025-1028.

Castle, J.W., and R.B. Miller. 2000. Recognition and hydrologic significance of passive-margin updip sequences: An example from Eocene coastal-plain deposits, USA. *Journal of Sedimentary Research,* 70:1290-1301.

Cey, B.D., S.L. Barbour, and M.J. Hendry. 2001. Osmotic flow through a Cretaceous clay in southern Saskatchewan, Canada. *Canadian Geotechnical Journal,* 38:1025-1033.

Chapman, S.W. and B.L. Parker. N.d. Plume persistence due to aquitard back-diffusion following DNAPL source removal or isolation, *Water Resources Research*, Forthcoming.

Chapron, C.D., N.A. Ballester, J.H. Fontaine, C.N. Frades, and A.B. Margolin. 2000. Detection of astroviruses, enteroviruses, and adenovirus types 40 and 41 in surface waters collected and evaluated by the information collection rule and an integrated cell culture-nested PCR procedure. *Applied and Environmental Microbiology,* 66:2520-2525.

Chattopadhyay, D., S. Chattopadhyay, W.G. Lyon, and J.T. Wilson. 2002. Effect of surfactants on the survival and sorption of viruses. *Environmental Science & Technology,* 36:4017-4024.

Chernyshev, S.N. and W.R. Dearman. 1991. *Rock Fractures*. London: Butterworth-Heinemann.

Cherry, J.A. 1983. Contaminant migration in groundwater with emphasis on hazardous waste disposal. In Proc. Groundwater Resources And Contamination In The United States, U.S. National Science Foundation PRA Report 83-12: 127-150.

Cherry, J.A. 1989. Hydrogeologic contaminant behaviour in fractured and unfractured clayey deposits in Canada. In *Proc.International Symposium on Contaminant Transport in Groundwater*, Edited by E. H. Kobus and W. Kinzelbach. Stuttgart, Germany: Balkema, A.A., Rotterdam: 11-20.

Cherry, J.A., B.T. Beswick, W.E. Clister, and M. Lutchman. 1971. Flow pattern and hydrochemistry of two shallow groundwater regimes in the Lake Agassiz basin, southern Manitoba. In *Geological Association of Canada Special Paper No.9* Edited by A.C. Turnock: 321-332..

Cherry, J.A., G.E. Grisak, and W.E. Clister. 1973. Hydrogeologic studies at a sub-surface radioactive-waste management site in west-central Canada. In *Underground waste management and artificial recharge.* Second International Symposium. Edited by J. Braunstein. New Orleans: AAPG-USGS:436-467

Cherry, J.A., C. Mase, R. Blackport, E.A. Sudicky, G.I. Thomson, R.W. Gillham, and K. Novakowski. 1989. Evaluation of contaminant velocity in low-permeability fractured shale. Ontario Ministry of the Environment.

Chown, J.C., B.H. Kueper, and D.B. McWhorter. 1997. The use of upward hydraulic gradients to arrest downward DNAPL migration in rock fractures. *Ground Water,* 35:483-491.

Chu, Y., and Y. Jin. 2001. Mechanisms of virus removal during transport in unsaturated porous media. *Water Resources Research,* 37:253-263.

Chu, Y., Y. Jin, and M.V. Yates. 2000. Virus transport through saturated sand columns as affected by different buffer solutions. *Journal of Environmental Quality,* 29:1103-1110.

Clark, I.D., and P. Fritz. 1997. *Environmental isotopes in hydrogeology*. Boca Raton, FL:CRC Press/Lewis Publishers.

Clauser, C. 1992. Permeability of Crystalline Rocks. *Eos* 73:233-240.

Clayton, L., and J.W. Attig. 1989. *Glacial Lake Wisconsin*. Boulder, CO:Geological Society of America.

Coleman, J.M., and S.M. Gagliano. 1965. Sedimentary structures: Mississippi River deltaic plain. *Primary Sedimentary Structures and Their Hydrodynamic Interpretation*, Special Pub. No.12, American Association of Petroleum Geologists: 133-148. Also published as: Technical Report No. 16, Part H, United States Coastal Gulf Studies, Louisiana State University, Baton Rouge, LA.

Conlon, T.D. 1998. *Hydrogeology and simulation of ground-water flow in the sandstone aquifer, northeastern Wisconsin*. Middleton, Wisconsin:.U.S. Geological Survey Water Resources Investigations Report 97-4096.

Cooke, M.L., and C.A. Underwood. 2001. Fracture termination and step-over at bedding interfaces due to frictional slip and interface opening. *Journal of Structural Geology,* 23:223-238.

Corbet, T.F. and C.M. Bethke. 1992. Disequilibrium fluid pressures ad groundwater flow in the western Canada sedimentary basin. *Journal of Geophysical Research*, 97(B5):7203-7217.

Corrigan, C.A., H.E. Jamieson, and V.H. Remenda. 2001. Fracture wall cements and coatings from two clayey till aquitards. *Ground Water,* 39:786-794.

Crattie, T.B. 1991. Delineation of a Discontinuous Aquitard with Vertical Electrical Soundings, San-Bernardino Valley, Southern California - Discussion. *Ground Water,* 29:924-924.

Craun, G.F., P.S. Berger, and R.L. Calderon. 1997. Coliform bacteria and waterborne disease outbreaks. *Journal American Water Works Association,* 89:96-104.

Crooks, V.E. and R.M. Quigley. 1984. Saline leachate migration through clay: A comparative laboratory and field investigation. *Canadian Geotechnical Journal*, 21: 349-362.

D'Amore, D.V., S.R. Stewart, J.H. Huddleston, and J.R. Glasmann. 2000. Stratigraphy and hydrology of the Jackson-Frazier wetland, Oregon. *Soil Science Society of America Journal,* 64:1535-1543.

Dansart, A.M. 2001. Effects of sand-filled ice-wedge casts on the spatial variability of unsaturated flow and mass transport. Master's thesis. University of Wisconsin, Madison.

D'Astous, A.Y., W.W. Ruland, J.R.G. Bruce, J.A. Cherry, and R.W. Gillham. 1989. Fracture effects in the shallow groundwater zone in weathered Sarnia-area clay. *Canadian Geotechnical Journal,* 26:43-56.

Davis, S.N. 1972. Use of naturally occurring phenomena to study hydraulic diffusivities of aquitards. *Water Resources Research,* 8:500-507.

Day, M.J. 1977. Analysis of movement and hydrochemistry of groundwater in fractured clay and till deposits of the Winnipeg area, Manitoba. Master's thesis. University of Waterloo Waterloo, Ontario.

Dershowitz, W., G. Lee, J. Geier, S. Hitchcock, and P. LaPointe. 1993. FRACMAN, Version 2.306: Interactive Discrete Feature Data Analysis, Geometric Modeling, and Exploration Simulation. Seattle, Washington: Golder Associates..

Desaulniers, D.E. 1980. Origin, age and movement of pore water in argillaceous Quaternary deposits at four sites in southwestern Ontario. Master's thesis. University of Waterloo, Waterloo, Ontario

Desaulniers, D.E. 1986. Groundwater origin, geochemistry and solute transport in three major glacial clay plains of east-central North America. Ph.D. diss., University of Waterloo, Waterloo, Ontario.

Desaulniers, D.E., and J.A. Cherry. 1988. Origin and movement of groundwater and major ions in a thick deposit of Champlain Sea clay near Montreal. *Canadian Geotechnical Journal,* 26:80-89.

Desaulniers, D.E., and J.A. Cherry. 1989. Origin and movement of groundwater and major ions in a thick deposit of Champlain Sea clay near Montreal. *Canadian Geotechnical Journal,* 26:80-89.

Desaulniers, D.E., J.A. Cherry, and P. Fritz. 1981. Origin, age and movement of pore water in argillaceous Quaternary deposits at four sites in southwestern Ontario. *Journal of Hydrology,* 50:231-257.

Desaulniers, D.E., R.S. Kaufmann, J.A. Cherry, and H.W. Bentley. 1986. 37Cl-35Cl variations in a diffusion-controlled groundwater system. *Geochimica et Cosmochimica Acta,* 50:1757-1764.

Desbarats, A.J., M.J. Hinton, C.E. Logan, and D.R. Sharpe. 2001. Geostatistical mapping of leakance in a regional aquitard, Oak Ridges Moraine area, Ontario, Canada. *Hydrogeology Journal,* 9:79-96.

Devlin, J.F., and B.L. Parker. 1996. Optimum hydraulic conductivity to limit contaminant flux through cutoff walls. *Ground Water,* 34:719-726.

Döll, P. and W. Schneider. 1995. Lab and Field-Measurements of the Hydraulic Conductivity of Clayey Silts. *Ground Water,* 33:884-891.

Domenico, P.A., and F.W. Schwartz. 1990. *Physical and Chemical Hydrogeology*. New York: John Wiley and Sons.

Dominic, D.F., R.W. Ritzi, and K.W. Kausch. 1996. Aquitard distribution in a northern reach of the Miami Valley Aquifer, Ohio, USA; Part 2, Interpretation of facies and geostatistical results. *Hydrogeology Journal,* 4:25-35.

Dowd, S.E., S.D. Pillai, S. Wang, and M.Y. Corapcioglu. 1998. Delineating the specific influence of virus isoelectric point and size on virus adsorption and transport through sandy soils. *Applied Environmental Microbiology,* 64:405-410.

Duboise, S.M., B.E. Moore, and B.P. Sagik. 1976. Poliovirus survival and movement in a sandy forest soil. *Applied Environmental Microbiology,* 31:536-543.

Dutton, A.R., E.W. Collins, S.D. Hovorka, R.E. Mace, B.R. Scanlon, and J. Xiang. 1994. Occurrence and movement of ground water in Austin Chalk and Eagle Ford and Ozan Formations at the Superconducting Super Collider (SSC) site, Ellis County, Texas. Topical Report prepared for Texas National Research Laboratory Commission Bureau of Economic Geology, Austin, Tex: The University of Texas at Austin.

Dutton, A.R., and W.W. Simpkins. 1989. Isotopic evidence for paleohydrologic evolution of ground-water flow paths, southern Great Plains, United States. *Geology,* 17:653-656.

Eaton, T.T. 2002. Fracture heterogeneity and hydrogeology of the Maquoketa aquitard, southeastern Wisconsin. Ph.D. diss. University of Wisconsin, Madison.

Eaton, T.T., M.P. Anderson, and K.R. Bradbury. 2001. Heterogeneity in groundwater flow modeling; simulating fractures in a regional aquitard. Geological Society of America, 2001 annual meeting. Boulder, CO: Geological Society of America (GSA).

Eaton, T.T., and K.R. Bradbury. 2003. Hydraulic transience and the role of bedding fractures in a bedrock aquitard, southeastern Wisconsin, USA. *Geophysical Research Letters,* 30(10): art no.-1961.

Eaton, T.T., K.R. Bradbury, M.P. Anderson, and H.F. Wang. 2000. Evaluation of the hydraulic properties of a sparsely fractured confining unit; importance of fracture connectivity. Geological Society of America 2000 annual meeting. Boulder, CO: Geological Society of America (GSA).

EddyDilek, C.A., B.B. Looney, P. Hoekstra, N. Harthill, M. Blohm, and D.R. Phillips. 1997. Definition of a critical confining zone using surface geophysical methods. *Ground Water,* 35:451-462.

Einarson, M.D. 2005. Multi-Level Ground Water Monitoring. In *Practical Handbook of Ground-Water Monitoring*. Edited by D.M. Nelson. Boca Ratan, FL: CRC Press, forthcoming.

Erskine, J.A., and A.T. Fisher. 2002. Implications of geometry, age and depositional history of a shallow confining layer in a coastal groundwater basin, Monterey County, California, USA. *Geophysical Research Letters* 29 (24):art. no.-2178.

Farah, E.A., B.L. Parker, and J.A. Cherry. 2003. The Use of Hydraulic Head and Atmospheric Tritium to Identify Presence of Fractures in Clayey Aquitards: Numerical Analysis. *in* Eos. Trans. AGU, 84(46), Fall Meet. Suppl., Abstract H21D-0865, 2003.

Farah, E.A., B.L. Parker, and J.A. Cherry. 2005. Sensitivity of Atmospheric Tritium for Identifying Fractures in Clayey Aquitards. In *Proc. of Groundwater Quality 2004, 4th International Conference: Bringing Groundwater Quality Research to the Watershed Scale*, Waterloo, ON: IAH, forthcoming.

Farvolden, R.N., and J.A. Cherry. 1988. St. Lawrence Lowland. Pages 133-140 In *The Geology of North America*. Edited by W. Back, J.S. Rosenshein, and P.R. Seaber. Boulder, CO: Geological Society of America.

Faulkner, B.R., W.G. Lyon, F.A. Khan, and S. Chattopadhyay. 2002. Predicting attenuation of viruses during percolation in soils. 1. Probabilistic model. EPA/600/R-02/051a, Cincinnati, OH: U.S. Environmental Protection Agency.

Feinstein, D.T., D.J. Hart, T.T. Eaton, J.T. Krohelski, and K.R. Bradbury. 2004. Simulation of regional groundwater flow in southeastern Wisconsin. Wisconsin Geological and Natural History Survey Open-File Report 2004-01. Madison, WI: WGNHS.

Ferris, J.G., D.B. Knowles, R.H. Brown and R.W. Stallman. 1962. *Theory of Aquifer Tests*. U.S. Geological Survey Water Supply Paper 1536-E. Washington, D.C.: USGS.

Fernandez, F., and R.M. Quigley. 1985. Hydraulic Conductivity of Natural Clays Permeated with Simple Liquid Hydrocarbons. *Canadian Geotechnical Journal,* 22:205-214.

Fernandez, F., and R.M. Quigley. 1988. Viscosity and Dielectric-Constant Controls on the Hydraulic Conductivity of Clayey Soils Permeated with Water-Soluble Organics. *Canadian Geotechnical Journal,* 25:582-589.

Fernandez, F., and R.M. Quigley. 1991. Controlling the Destructive Effects of Clay Organic Liquid Interactions by Application of Effective Stresses. *Canadian Geotechnical Journal,* 28:388-398.

Fetter, C.W., Jr. 2001. *Applied Hydrogeology*, 4th edition. Upper Saddle River, NJ:. Prentice-Hall.

Flint, A.L., L.E. Flint, E.M. Kwicklis, G.S. Bodvarsson, and J.M. Fabryka-Martin. 2001. Hydrology of Yucca Mountain, Nevada. *Reviews of Geophysics,* 39:447-470.

Fogg, G.E. 1990. Architecture and interconnectedness of geologic media: role of the low-permeability facies in flow and transport. In *Hydrogeology of Low Permeability Environments.* Edited by S.P. Neuman, I. Neretnieks, and H. Heise, Hanover: 19-40.

Fortin, G., G. van der Kamp, and J.A. Cherry. 1991. Hydrogeology and Hydrochemistry of an Aquifer Aquitard System within Glacial Deposits, Saskatchewan, Canada. *Journal of Hydrology,* 126:265-292.

Foster, S.S.D. 1975. The Chalk groundwater tritium anomaly; a possible explanation. *Journal of Hydrology,* 25:159-165.

Fout, G.S., B.C. Martinson, M.W.N. Moyer, and D.R. Dahling. 2003. A multiplex reverse transcription-PCR method for detection of human enteric viruses in groundwater. *Applied and Environmental Microbiology,* 69:3158-3164.

Fredericia, J. 1990. Saturated Hydraulic Conductivity of Clayey Tills and the Role of Fractures. *Nordic Hydrology,* 21:119-132.

Freeze, R.A., and J.A. Cherry. 1979. *Groundwater*. Englewood Cliffs, NJ: Prentice-Hall.

Freeze, R.A., and D.B. McWhorter. 1997. A framework for assessing risk reduction due to DNAPL mass removal from low-permeability soils. *Ground Water,* 35:111-123.

Freeze, R.A., and P.A. Witherspoon. 1967. Theoretical analysis of regional groundwater flow; [Part] 2, Effect of. *Water Resources Research,* 3:623-634.

Gautschi, A. 2001. Hydrogeology of a fractured shale (Opalinus Clay): Implications for deep geological disposal of radioactive wastes. *Hydrogeology Journal,* 9:97-107.

Geistlinger, H., J. Dermietzel, D. Lazik, and W. Glaesser. 1998. Analytical modelling of diffusion-limited contaminant release from the Bitterfeld lignite seam aquitard. In *Proc. of International Conference on Groundwater Quality, Remediation and Protection,* Tübingen, Germany. Edited by K. Kovar and M. Herbert. Wallingford, Oxfordshire: International Association of Hydrological Sciences: 591-594.

Gerba, C.P., and G. Bitton. 1984. Microbial pollutants; their survival and transport pattern to groundwater. In Groundwater pollution microbiology. Edited by G. Bitton and P. Gerba Charles, editors. New York, NY: John Wiley & Sons: 65-88.

Gerba, C.P., and J.B. Rose. 1990. Viruses in source and drinking water. In *Drinking Water Microbiology. Progress and Recent Developments*. Edited by G. A. McFeters. New York: Springer-Verlag: 380-396.

Gerber, R.E., J.I. Boyce, and K.W.F. Howard. 2001. Evaluation of heterogeneity and field-scale groundwater flow regime in a leaky till aquitard. *Hydrogeology Journal,* 9:60-78.

Gerber, R.E., and K.W.F. Howard. 1996. Evidence for recent groundwater flow through late Wisconsinan till near Toronto, Ontario. *Canadian Geotechnical Journal,* 33:538-555.

Gillham, R.W., and J.A. Cherry. 1982. Contaminant migration in saturated unconsolidated geologic deposits. In *Recent Trends in Hydrogeology*, Geological Society of America Special Publication 189. Edited by N.P. Narasimhan. Berkeley, CA: Geological Society of America: 31-62.

Goodall, D.C., and R.M. Quigley. 1977. Pollutant migration from two sanitary landfill sites near Sarnia, Ontario. *Canadian Geotechnical Journal,* 14:223-236.

Gordon, C., and S. Toze. 2003. Influence of groundwater characteristics on the survival of enteric viruses. *Journal of Applied Microbiology,* 95:536-544.

Goyal, M.S., and C.P. Gerba. 1979. Comparative adsorption of human enteroviruses, simian rotavirus, and selected bacteriophages to soils. *Appl Environ Microbiol,* 38:241-247.

Graese, A.M., R.A. Bauer, B.B. Curry, R.C. Vaiden, W.G. Dixon, and J.P. Kempton. 1988. Geological-geotechnical studies for citing the Superconducting Super Collider in Illinois: regional summary. Champaign, IL: Illinois State Geological Survey, Environmental Geology Notes 123.

Green, W.J., G.F. Lee, and R.A. Jones. 1981. Clay-soils permeability and hazardous waste storage. *Journal - Water Pollution Control Federation,* 53:1347-1354.

Griffin, R.A., R.E. Hughes, L.R. Follmer, C.J. Stohr, W.J. Morse, T.M. Johnson, J.K. Bartz, J.D. Steele, K. Cartwright, M.M. Killey, and P.B. Dumontelle. 1984. Migration of industrial chemicals and soil-waste interactions at Wilsonville, Illinois. In Proc. of Land Disposal of Hazardous Waste, Tenth Annual Research Symposium. EPA-600/9-84-007, Ft. Mitchell, KY: 61-77.

Grisak, G.E., and J.A. Cherry. 1975. Hydrologic characteristics and response of fractured till and clay confining a shallow aquifer. *Canadian Geotechnical Journal,* 12:23-43.

Grisak, G.E., J.A. Cherry, J.A. Vonhof, and J.P. Blumele. 1976. Hydrogeologic and hydrochemical properties of fractured till in the interior plains region. Edited by R. F. Legget, editor. Royal Society of Canada, Ottawa. 304-335.

Grisak, G.E., and J.F. Pickens. 1980. Solute transport through fractured media, Part 1: The effect of matrix diffusion. Water Resources Research, 16:719-730.

Grisak, G.E., J.F. Pickens, and J.A. Cherry. 1980. Solution transport through fractured media, Part 2, Column study of fractured till. *Water Resources Research,* 16:731-739.

Gross, M.R., M.P. Fischer, T. Engelder, and R.J. Greenfield. 1995. Factors controlling joint spacing in interbedded sedimentary rocks; integrating numerical models with field observations from the Monterey Formation, USA. In *Fractography; fracture topography as a tool in fracture mechanics and stress analysis.* Edited by S. Ameen Mohammed. London, U.K.: Geological Society of London: 215-233

Guan, H.D., D. Schulze-Makuch, S. Schaffer, and S.D. Pillai. 2003. The effect of critical pH on virus fate and transport in saturated porous medium. *Ground Water,* 41:701-708.

Haefner, R.J. 2000. Characterization methods for fractured glacial tills. *Ohio Journal of Science,* 100:73-87.

Haldorsen, S., and J. Kruger. 1990. Till Genesis and Hydrogeological Properties. *Nordic Hydrology,* 21:81-94.

Hanor, J.S. 1993. Effective Hydraulic Conductivity of Fractured Clay Beds at a Hazardous-Waste Landfill, Louisiana Gulf-Coast. *Water Resources Research,* 29:3691-3698.

Harloff, N.C. 1942. Lacustrine clays of Glacial Lake Wisconsin as determined by fire protection well records. Bachelor of Philosophy diss. University of Wisconsin, Madison, WI.

Harrison, B., E.A. Sudicky, and J.A. Cherry. 1992. Numerical-Analysis of Solute Migration through Fractured Clayey Deposits into Underlying Aquifers. *Water Resources Research,* 28:515-526.

Hart, D.J., K.R. Bradbury, and D.T. Feinstein. [N.d.] The vertical hydraulic conductivity of an aquitard: and evaluation of the Maquoketa Formation at two spatial scales *Ground Water*, Forthcoming.

Hearn, M.R. 1996. Stratigraphic and diagenetic controls on aquitard integrity and hydrocarbon entrapment, Bashaw Reef Complex, Alberta, Canada. Master's thesis, Edmonton, Alta: University of Alberta.

Helgeson, D.E., and A. Aydin. 1991. Characteristics of Joint Propagation across Layer Interfaces in Sedimentary-Rocks. *Journal of Structural Geology,* 13:897-911.

Helmke, M.F., W.W. Simpkins, and R. Horton. 1996. A multi-disciplinary approach for estimating effective porosity from large, undisturbed till cores. Proc. Of *Geological Society of America 28th annual meeting.* Boulder, CO: Geological Society of America.

Helmke, M F., W.W. Simpkins, and R. Horton. 1997. Evaluation of solute transport through fractured till using dual porosity and discrete fracture models. Proc. of *Geological Society of America, North-Central Section, 31st annual meeting.* Boulder, CO: Geological Society of America (GSA).

Helmke, M.F., W.W. Simpkins, and R. Horton. 1999. Use of multiple tracers and measured effective diffusion coefficients to constrain solute transport models of fractured till. Proc. of *Geological Society of America 1999 annual meeting.* Boulder, CO: Geological Society of America.

Hendry, M.J. 1982. Hydraulic Conductivity of a Glacial Till in Alberta. *Ground Water,* 20:162-169.

Hendry, M.J. 1988. Ground water movement in clayey till in the prairie region. *Ground Water,* 26(5): 607-614.

Hendry, M.J., and L.I. Wassenaar. 1997. Natural groundwater tracers (18O, 2H, Cl, 37Cl/ 36Cl) in a thick aquitard system provide information on the geologic and climatic history of southern Saskatchewan, Canada. Proc. Of *Geological Society of America 1997 annual meeting*. Boulder, CO: Geological Society of America.

Hendry, M.J., and L.I. Wassenaar. 1999. Implications of the distribution of delta D in pore waters for groundwater flow and the timing of geologic events in a thick aquitard system. *Water Resources Research,* 35:1751-1760.

Hendry, M.J., and L.I. Wassenaar. 2005. The origin and long-term transport of DOC in a clay-rich aquitard. *Water Resources Research*, forthcoming.

Hendry, M.J., L.I. Wassenaar, and T. Kotzer. 2000. Chloride and chlorine isotopes (Cl-36 and delta Cl-37) as tracers of solute migration in a thick, clay-rich aquitard system. *Water Resources Research,* 36(1):285-296.

Herzog, B.L., R.A. Griffin, C.J. Stohr, L.R. Follmer, W.J. Morse, and W.J. Su. 1989. Investigation of failure mechanisms and migration of organic chemicals at Wilsonville, Illinois. *Ground Water Monitoring Review,* 9:82-89.

Hinsby, K., L.D. McKay, P. Jorgensen, M. Lenczewski, and C.P. Gerba. 1996. Fracture aperture measurements and migration of solutes, viruses, and immiscible creosote in a column of clay-rich till. *Ground Water,* 34:1065-1075.

Hornung, J., and T. Aigner. 1999. Reservoir and aquifer characterization of fluvial architectural elements: Stubensandstein, Upper Triassic, southwest Germany. *Sedimentary Geology,* 129:215-280.

Hsieh, P.A. 1998. Scale effects in fluid flow through fractured geologic media. In *Scale Dependence and Scale Invariance in Hydrology*. Edited by G. Sposito. Cambridge, U.K.: Cambridge University Press: 335-353.

Hunt, C. D., Jr., C. J. Ewart, and C. I. Voss. 1988. Region 27, Hawaiian Islands. In *The Geology of North America: Hydrogeology* Edited by W. Back, S. Rosenshein Joseph, and R. Seaber Paul. Boulder, CO:. Geological Society of America: 255-262.

Hurst, C.J., C.P. Gerba, and I. Cech. 1980. Effects of environmental variables and soil characteristics on virus survival in soil. *Appl Environ Microbiol,* 40:1067-1079.

Hurst, C.J., C.P. Gerba, J.C. Lance, and R.C. Rice. 1980. Survival of enteroviruses in rapid-infiltration basins during the land application of wastewater. *Appl Environ Microbiol,* 40:192-200.

Husain, M.M. 1996. Origin and persistence of Pleistocene and Holocene water in a regional clayey aquitard and underlying aquifer in part of southwestern Ontario. PhD. diss., University of Waterloo, Waterloo, Ontario.

Husain, M.M., J.A. Cherry, S. Fidler, and S.K. Frape. 1998. On the long-term hydraulic gradient in the thick clayey aquitard in the Sarnia region, Ontario. *Canadian Geotechnical Journal,* 35:986-1003.

Husain, M.M., J.A. Cherry, and S.K. Frape. 2004. The persistence of a large stagnation zone in a developed regional aquifer, southwestern Ontario. *Canadian Geotechnical Journal, 41:943-958.*

James, B.R., and R.A. Freeze. 1993. The Worth of Data in Predicting Aquitard Continuity in Hydrogeological Design. *Water Resources Research,* 29:2049-2065.

Jancin, M., and D.D. Clark. 1993. Subsidence-Sinkhole Development in Light of Mud Infiltrate Structures within Interstriatal Karst of the Coastal-Plain, Southeast United-States. *Environmental Geology,* 22:330-336.

Jin, Y., E. Pratt, and M.V. Yates. 2000. Effect of Mineral Colloids on Virus Transport through Saturated Sand Columns. *Journal of Environmental Quality,* 29:532-539.

Johnson, R.L., J.A. Cherry, and J.F. Pankow. 1989. Diffusive contaminant transport in natural clay; a field example and implications for clay-lined waste disposal sites. *Environmental Science and Technology,* 23:340-349.

Jones, L., T. Lemar, and C.T. Tsai. 1992. Results of 2 Pumping Tests in Wisconsin Age Weathered Till in Iowa. *Ground Water,* 30:529-538.

Jorgensen, D.G., J. Downey, A.R. Dutton, and R.W. Maclay. 1988. Region 16, Central Nonglaciated Plains. In *The Geology of North America: Hydrogeology*. Edited by W. Back, J.S. Rosenshein, and P.R. Seaber. Boulder, CO: Geological Society of America: 141-156.

Jorgensen, P.R., K. Broholm, T.O. Sonnenborg, and E. Arvin. 1998. DNAPL transport through macroporous, clayey till columns. *Ground Water,* 36:651-660.

Jorgensen, P.R., and J. Fredericia. 1992. Migration of Nutrients, Pesticides and Heavy-Metals in Fractured Clayey Till. *Geotechnique,* 42:67-77.

Jorgensen, P.R., T. Helstrup, J. Urup, and D. Seifert. 2004a. Modeling of non-reactive solute transport in fractured clayey till during variable flow rate and time. *Journal of Contaminant Hydrology,* 68(3-4):193-216.

Jorgensen, P.R., M. Hoffmann, J. Histrup, C. Bryde, R. Bossi, and K.G. Villholt. 2002. Preferential flow and pesticide transport in a clay-rich till: Field, laboratory and modeling analyses. *Water Resources Research*, 38(11): 2801-2815.

Jorgensen, P.R., K.E.K. Klint, and J.P. Kistrup. 2003. Monitoring well interception with fractures in clayey till. *Ground Water*, 41(6): 772-779.

Jorgensen, P.R., L.D. Mackay, and N.ZH. Spliid. 1998. Evaluation of chloride and pesticide transport in a fractured clayey till using large undisturbed columns and numerical modeling. *Water Resources Research*, 34(2): 539-553.

Jorgensen, P.R., L.D. McKay, and J.P. Kistrup. 2004. Aquifer vulnerability to pesticide migration through till aquitards. *Ground Water*, 42(6): 841-855.

Jorgensen, P.R., J. Urup, T. Helstrup, M.B. Jensen, F. Eiland, and F.P. Vinther. 2004b. Transport and removal of nitrate in clayey till underneath forest and arable land. *Journal of Contaminant Hydrology*, 73:270-286.

Keller, C.K., G. van der Kamp, and J.A. Cherry. 1986. Fracture Permeability and Groundwater-Flow in Clayey Till near Saskatoon, Saskatchewan. *Canadian Geotechnical Journal,* 23:229-240.

Keller, C.K., G. van der Kamp, and J.A. Cherry. 1989. Multiscale Study of the Permeability of a Thick Clayey Till. *Water Resources Research,* 25:2299-2317.

Kelln, C.J., L.I. Wassenaar, and M.J. Hendry. 2001. Stable isotope (delta 18O, delta D) of pore waters in clay-rich aquitards: A comparison and evaluation of measurement techniques. *Ground Water Monitoring and Remediation*, 21(2): 108-116.

Kepferle, R.C., P.E. Potter, and W.A. Pryor. 1981. *Stratigraphy of the Chattanooga Shale (Upper Devonian and Lower Mississippian) in vicinity of Big Stone Gap, Wise County, Virginia, with a section* on Petrology, by T.V. Stenbeck: U.S. Geological Survey Bulletin 1499.

Keswick, B.H., and C.P. Gerba. 1980. Viruses in groundwater. *Environmental Science & Technology,* 14:1290-1297.

King, K.S., R.M. Quigley, F. Fernandez, D.W. Reades, and A. Bacopoulos. 1993. Hydraulic Conductivity and Diffusion Monitoring of the Keele Valley Landfill Liner, Maple, Ontario. *Canadian Geotechnical Journal,* 30:124-134.

Kinoshita, T., R.C. Bales, K.M. Maguire, and C.P. Gerba. 1993. Effect of Ph on Bacteriophage Transport through Sandy Soils. *Journal of Contaminant Hydrology,* 14:55-70.

Kirkpatrick, G.A. 1998. Infiltration, migration and phase transfer of common DNAPLs in a naturally fractured clayey aquitard: Field experiments. Master's thesis. University of Waterloo, Waterloo, Ontario.

Klint, K.E.S. 2001. Fractures in Glacigene Diamict Deposits; Origin and Distribution. Ph.D. diss. University of Copenhagen.

Klint, K.E.S., and P. Gravesen. 1999. Fractures and biopores in Weichselian clayey till aquitards at Flakkebjerg, Denmark. *Nordic Hydrology,* 30:267-284.

Kolm, K.E., and K.D. Peter. 1984. A Possible Relation Between Lineaments and Leakage Through Confining Layers in South Dakota. In *Proc. of Geohydrology Dakota Aquifer Symposium.* Edited by D.G. Jorgensen and D.C. Signor.. Lincoln, NE: The National Water Well Association: 98-111

Kreitler, C.W., R.S. Fisher, R.K. Senger, S.D. Hovorka, and A.R. Dutton. 1985. Hydrology of an evaporite aquitard: Permian evaporite strata, Palo Duro basin, Texas. In *Hydrology of Rocks of Low Permeability.* Edited by E.S. Simpson. Tucson, AZ:International Association of Hydrogeologists: 150-168.

Krohelski, J.T., K.R. Bradbury, R.J. Hunt, and S.K. Swanson. 2000. Numerical simulation of groundwater flow in Dane County, Wisconsin. Madison, Wisc.: Wisconsin Geological and Natural History Survey.

Kueper, B.H., and D.B. McWhorter. 1991. The Behavior of Dense, Nonaqueous Phase Liquids in Fractured Clay and Rock. *Ground Water,* 29:716-728.

Lacombe, S., E.A. Sudicky, S.K. Frape, and A.J.A. Unger. 1995. Influence of Leaky Boreholes on Cross-Formational Groundwater-Flow and Contaminant Transport. *Water Resources Research,* 31:1871-1882.

Lafleur, J., F. Giroux, and M. Huot. 1987. Field permeability of the weathered Champlain clay crust. *Canadian Geotechnical Journal,* 24:581-589.

Lance, J.C., and C.P. Gerba. 1984a. Virus movement in soil during saturated and unsaturated flow. *Appl Environ Microbiol,* 47:335-337.

Lance, J.C., and C.P. Gerba. 1984b. Effect of ionic composition of suspending solution on virus adsorption by a soil column. *Appl Environ Microbiol,* 47:484-488.

Langevin, C.D. 2003. Stochastic ground water flow simulation with a fracture zone continuum model. *Ground Water,* 41:728-728.

Larson, K.R., C.K. Keller, P.B. Larson, and R.M. Allen-King. 2000. Water resource implications of O-18 and 2H distributions in a basalt aquifer system. *Ground Water,* 38:947-953.

Lawrence, J.R., M.J. Hendry, L.I. Wassenaar, G.M. Wolfaardt, J.J. Germida, and C.W. Greer. 2000. Distribution and biogeochemical importance of bacterial populations in a thick clay-rich aquitard system. *Microbial Ecology,* 40:273-291.

Lebbe, L., and N. Van Meir. 2000. Hydraulic conductivities of low permeability sediments inferred from a triple pumping test and observed vertical gradients. *Ground Water,* 38:76-88.

Lee, S.H., D.A. Levy, G.F. Craun, M.J. Beach, and R.L. Calderon. 2002. Surveillance for waterborne-disease outbreaks - United States, 1999-2000. *MMWR CDC Surveillance Summaries* 51:1-28.

Lefler, E., and Y. Kott. 1974. Viruses retention and survival in sand. In *Virus Survival in Water and Wastewater Systems*. Edited by J.F. Malina and B.P. Sagik. Austin, TX: Center for Research in Water Resources:84.

Leonhart, L.S., and D.R. Hargis. 1990. The hydrogeologic character of sedimentary interbeds in flood basalts. In *International Conference on Groundwater in Large Sedimentary Basins*. Edited by H. Ventriss.. Perth, Australia: Australian Water Resources Council: 90-99.

Lerner, D.N., M.W. Burston, and P.K. Bishop. 1993. Hydrogeology of the Coventry Region (UK) - an Urbanized, Multilayer, Dual-Porosity Aquifer System. *Journal of Hydrology,* 149:111-135.

Li, H.L., and J.J. Jiao. 2001a. Analytical studies of groundwater-head fluctuation in a coastal confined aquifer overlain by a semi-permeable layer with storage. *Advances in Water Resources,* 24:565-573.

Li, H.L., and J.J. Jiao. 2001b. Tide-induced groundwater fluctuation in a coastal leaky confined aquifer system extending under the sea. *Water Resources Research,* 37:1165-1171.

Li, T. 1991. Hydrogeologic characterization of a multiple aquifer fractured basalt system. PhD. Diss. University of Idaho.

Lin, W. 1978. Measuring the permeability of Eleana Argillite from area 17, Nevada Test Site, using the transient method. Rep. UCRL-52604. Livermore, Ca: Lawrence Livermore Lab.

Lind, B. B., and L. Lundin. 1990. Saturated Hydraulic Conductivity of Scandinavian Tills. *Nordic Hydrology,* 21:107-118.

Lindgren, R.J. 2001. *Ground-Water Recharge and Flowpaths near the edge of the Decorah-Platteville-Glenwood Confining Unit, Rochester, Minnesota*. U.S. Geological Survey Water Resources Investigations Report 00-4215.

Lindholm, G.F., and J.J. Vaccaro. 1988. Region 2, Columbia Lava Plateau. The Geology of North America : Hydrogeology. Edited by W. Back, S. Rosenshein Joseph, and R. Seaber Paul. Boulder, CO: Geological Society of America: 37-50.

Lissey, A. 1962. *Groundwater resources of the Regina area*. Regina, Sask.: Regina Engineering Department.

Liu, C.X. 1999. Analytical modeling of aquitard effects on contaminant transport and groundwater remediation. PhD. diss., The Johns Hopkins University, Baltimore, MD.

Liu, C.X., and W.P. Ball. 1999. Application of inverse methods to contaminant source identification from aquitard diffusion profiles at Dover AFB, Delaware. *Water Resources Research,* 35:1975-1985.

Liu, C.X., and W.P. Ball. 2002. Back diffusion of chlorinated solvent contaminants from a natural aquitard to a remediated aquifer under well-controlled field conditions: Predictions and measurements. *Ground Water,* 40:175-184.

Macfarlane, P.A., J.H. Doveton, H.R. Feldman, J.J. Butler, J.M. Combes, and D.R. Collins. 1994. Aquifer/Aquitard Units of the Dakota Aquifer System in Kansas - Methods of Delineation and Sedimentary Architecture Effects on Groundwater-Flow and Flow Properties. *Journal of Sedimentary Research Section B-Stratigraphy and Global Studies,* 64:464-480.

Maguire, T.J. 1998. Hydrogeologic aspects of the mechanical stratigraphy of the Newark Basin, a continental rift basin. *Northeastern Geology and Environmental Sciences,* 20:325-344.

Manzano, M., E. Custodio, and B.F. Jones. 1990. Progress in the understanding of groundwater flow through the aquitard of the Llobregat Delta; Barcelona, Spain. In *Livro de homenagem a Carlos Romariz.* Edited by M. Alves Carlos Alberto de and R. Fernandez Rubio. Lisbon, Portugal: Univ. Lisboa, Secc. Geol. Econ. Apl: 115-126

Martin, P.J., and E.O. Frind. 1998. Modeling a complex multi-aquifer system: The Waterloo Moraine. *Ground Water,* 36:679-690.

Marty, B., S. Dewonck, and C. France-Lanord. 2003. Geochemical evidence for efficient aquifer isolation over geologic timeframes. *Nature,* 425(4 Sept 03):55-58.

Mase, C., G. Thompson, J.A. Cherry, R. Blackport, and L. Smith. 1987. *Evaluation of contaminant velocity in low-permeability fractured shale.* Toronto, ON: Ontario Ministry of the Environment.

Mas-Pla, J., J. Bach, E. Vinals, J. Trilla, and J. Estralrich. 1999. Salinization Processes in a Coastal Leaky Aquifer System (Alt Emporda, NE Spain). *Physics and Chemistry of the Earth, Part B: Hydrology, Oceans and Atmosphere,* 24:337-341.

Mazor, E. 2003. *Chemical and isotopic groundwater hydrology,* 3rd edition. New York: Marcel Dekker.

Mazurek, M., G.W. Lanyon, S. Vomvoris, and A. Gautschi. 1998. Derivation and application of a geologic dataset for flow modelling by discrete fracture networks in low-permeability argillaceous rocks. *Journal of Contaminant Hydrology,* 35:1-17.

McDonald, M.G., and A.W. Harbaugh. 1988. *A modular three-dimensional finite-difference ground-water flow model.* Reston, Va.: U.S. Geological Survey..

McIelwain, T.A., D.W. Jackman, and P. Beukman. 1989. Characterization and remedial assessment of DNAPL PCB oil in fractured bedrock: a case study of the Smithville, Ontario site. In *Proc. of FOCUS Conference on Eastern Regional Groundwater Issues.* Kitchener, ON: National Well Water Association: 303-315

McKay, L.D., D.J. Balfour, and J.A. Cherry. 1998. Lateral chloride migration from a landfill in a fractured clay-rich glacial deposit. *Ground Water,* 36:988-999.

McKay, L.D., J.A. Cherry, R.C. Bales, M.T. Yahya, and C.P. Gerba. 1993. A field example of bacteriophage as tracers of fracture flow. *Environmental Science & Technology,* 27:1075-1079.

McKay, L.D., J.A. Cherry, and R.W. Gillham. 1993. Field Experiments in a Fractured Clay Till .1. Hydraulic Conductivity and Fracture Aperture. *Water Resources Research,* 29:1149-1162.

McKay, L.D., and J. Fredericia. 1995. Distribution, origin, and hydraulic influence of fractures in a clay-rich glacial deposit. *Canadian Geotechnical Journal,* 32:957-975.

McKay, L.D., R.W. Gillham, and J.A. Cherry. 1993. Field Experiments in a Fractured Clay Till .2. Solute and Colloid Transport. *Water Resources Research,* 29:3879-3890.

McKay, L.D., W.E. Sanford, and J.M. Strong. 2000. Field-scale migration of colloidal tracers in a fractured shale saprolite. *Ground Water,* 38:139-147.

McMahon, P.B. 2001. Aquifer/aquitard interfaces: mixing zones that enhance biogeochemical reactions. *Hydrogeology Journal,* 9:34-43.

McMahon, P.B., and F.H. Chapelle. 1991. Microbial-Production of Organic-Acids in Aquitard Sediments and Its Role in Aquifer Geochemistry. *Nature,* 349:233-235.

McWhorter, D.B., and B.H. Kueper. 1996. Mechanics and mathematics of the movement of dense non-aqueous phase liquids (DNAPLs) in porous media. In *Dense Chlorinated Solvents and Other DNAPLs in Groundwater; History, Behavior, and Remediation.* Edited by J.F. Pankow and J.A. Cherry. Rockwood, ON: Waterloo Press: 89-128.

Meldrum, C.I. 1999. Use of a new multilevel monitoring system for determining hydraulic head distribution in the Borden aquitard. Bachelor thesis, University of Waterloo, Waterloo, Ontario.

Meyboom, P. 1966. Unsteady groundwater flow near a willow ring in hummocky moraine. *Journal of Hydrology,* 4:38-62.

Michalski, A., and R. Britton. 1997. The role of bedding fractures in the hydrogeology of sedimentary bedrock - Evidence from the Newark Basin, New Jersey. *Ground Water,* 35:318-327.

Middleton, T.A., and J.A. Cherry. 1996. The effects of chlorinated solvents on the permeability of clays. In *Dense Chlorinated Solvents and Other DNAPLs in Groundwater; History, Behavior, and Remediation.* Edited by J.F. Pankow and J.A. Cherry. Rockwood, ON: Waterloo Press:. 313-335

Morrison, W.E.M., B.L. Parker, and J.A. Cherry. 1998. Hydrogeological controls in flow and fate of PCE DNAPL in a fractured and layered clayey aquitard; a Borden experiment. In *Proc. of Geological Society of America 1998 annual meeting.* Boulder, CO: Geological Society of America: United States. 174

Motz, L.H., G.D. Sousa, and M.D. Annable. 2001. Water budget and vertical conductance for Lowry (Sand Hill) Lake in north-central Florida, USA. *Journal of Hydrology,* 250:134-148.

Narasimhan, T.N. 2004. Fick's insights on liquid diffusion. *EOS*, Trans. Americal Geophysical Union 85(47): 499-502. November 23, 2004.

Narr, W., and J. Suppe. 1991. Joint Spacing in Sedimentary-Rocks. *Journal of Structural Geology,* 13:1037-48.

National Research Council. 1993. *Ground Water Vulnerability Assessment: Predicting Relative Contamination Potential Under Conditions of Uncertainty. Committee for Assessing Ground Water Vulnerability*. The National Academies Press.

Nativ, R., O. Dahan, I. Nissim, E. Adar, and M.A. Geyh. 1995. Questions pertaining to isotopic and hydrochemical evidence of fracture-controlled water flow and solute transport in a chalk aquitard. International-Association-of-Hydrological-Sciences Publication No.232: Wallingford, England: IAHS: 317-327

Nativ, R., A. Halleran, and A. Hunley. 1997. Evidence for ground-water circulation in the brine-filled aquitard, Oak Ridge, Tennessee. *Ground Water,* 35:647-659.

Nativ, R., and I. Nissim. 1992. Characterization of a Desert Aquitard - Hydrologic and Hydrochemical Considerations. *Ground Water,* 30:598-606.

Nelson, E.P., A.J. Kullman, M.H. Gardner, and M. Batzle. 1999. Fault-fracture networks and related fluid flow and sealing, Brushy Canyon Formation, West Texas. In *Faults and subsurface fluid flow in the shallow crust.* Edited by C. Haneberg William, S. Mozley Peter, J.C. Moore, and B. Goodwin Laurel. Washington, D.C.: American Geophysical Union. 69-81.

Neuman, S.P., and D.A. Gardner. 1989. Determination of aquitard/aquiclude hydraulic properties from arbitrary water-level fluctuations by deconvolution. *Ground Water,* 27:66-76.

Neuman, S.P., and I. Neretnieks, editors. 1990. *Hydrogeology of low permeability environments*. Hanover, Federal Republic of Germany: Verlag Heinz Heise.

Neuman, S.P., and P.A. Witherspoon. 1972. Field determination of the hydraulic parameters of leaky multiple aquifer systems. *Water Resources Research,* 8:1284-1298.

Neuzil, C.E. 1986. Groundwater flow in low-permeability environments. *Water Resources Research,* 22:1163-1195.

Neuzil, C.E. 1988. Groundwater-Flow in a Cretaceous Shale. *Ground Water,* 26:784-784.

Neuzil, C.E. 1993. Low Fluid Pressure within the Pierre Shale - a Transient-Response to Erosion. *Water Resources Research,* 29:2007-2020.

Neuzil, C.E. 1994. How Permeable Are Clays And Shales? *Water Resources Research,* 30:145-150.

Neuzil, C.E. 1995. Abnormal pressures as hydrodynamic phenomena. *American Journal of Science,* 295:742-786.

Neuzil, C.E. 2000. Osmotic generation of 'anomalous' fluid pressures in geological environments. *Nature,* 403:182-184.

Neuzil, C.E. 2002. Decreases in groundwater storage: the problem of confining layers. In *Proc. of Geological Society of America 2002 Annual Meeting*. Denver, CO: Geological Society of America.

Neuzil, C.E., J.D. Bredehoeft, and R.G. Wolff. 1984. Leakage and Fracture Permeability in the Cretaceous Shales Confining the Dakota Aquifer in South Dakota. In *Geohydrology Dakota Aquifer Symposium. Edited by* D.G. Jorgensen and D.C. Signor.. Lincoln, NE: The National Water Well Association: 133-140

Nilsson, B., R.C. Sidle, K.E. Klint, C.E. Boggild, and K. Broholm. 2001. Mass transport and scale-dependent hydraulic tests in a heterogeneous glacial till-sandy aquifer system. *Journal of Hydrology,* 243:162-179.

Norris, S.E. 1959. *Vertical leakage through till as a source of recharge to a buried-valley aquifer at Dayton, Ohio*. Columbus, Ohio: Department of Natural Resources.

Novakowski, K.S., and G.S. Bickerton. 1997. Borehole measurement of the hydraulic properties of low-permeability rock. *Water Resources Research,* 33:2509-2517.

Oatfield, W.J., and J.B. Czarnecki. 1991. Hydrogeologic Inferences from Drillers Logs and from Gravity and Resistivity Surveys in the Amargosa Desert, Southern Nevada. *Journal of Hydrology,* 124:131-158.

O'Hara, S.K., B.L. Parker, P.R. Jorgensen, and J.A. Cherry. 2000. Trichloroethene DNAPL flow and mass distribution in naturally fractured clay: Evidence of aperture variability. *Water Resources Research,* 36:135-147.

Oliveira, D.P., and N. Sitar. 1985. Ground Water Contamination from Underground Solvent Storage Tanks. In *Proc. of the Fifth National Symposium and Exposition on Aquifer Restoration and Ground Water Monitoring*, Columbus, Ohio: 691-708.

Ortega-Guerrero, A., J.A. Cherry, and R.O. Aravena. 1997. Origin of pore water and salinity in the lacustrine aquitard overlying the regional aquifer of Mexico City. *Journal of Hydrology,* 197:47-69.

Ortega-Guerrero, A., J.A. Cherry, and D.L. Rudolph. 1993. Large-Scale Aquitard Consolidation near Mexico-City. *Ground Water,* 31:708-718.

Ortega-Guerrero, M.A. 1993. Origin and migration of pore water and salinity in the consolidating Chalco aquitard, near Mexico City. PhD. diss. University of Waterloo, Waterloo, Ontario.

Ortega-Guerrero, M.A. and R.N. Farvolden. 1989. Computer analysis of regional groundwater flow and boundary conditions in the Basin of Mexico. *Journal of Hydrology,* 110:271-294.

Ortega-Guerrero, A., D.L. Rudolph, and J.A. Cherry. 1999. Analysis of long-term land subsidence near Mexico City: Field investigations and predictive modeling. *Water Resources Research,* 35:3327-3341.

Owen, W.P., S.K. Park, and T.C. Lee. 1991. Delineation of a Discontinuous Aquitard with Vertical Electrical Soundings, San-Bernardino-Valley, Southern California. *Ground Water,* 29:418-424.

Pankow, J.F., and J.A. Cherry. 1996. *Dense Chlorinated Solvents and Other DNAPLs in Groundwater : History, Behavior, andRemediation.* Rockwood, ON: Waterloo Press.

Pankow, J.F., R.L. Johnson, J.P. Hewetson, and J.A. Cherry. 1986. An evaluation of contaminant migration patterns at two waste disposal sites on fractured porous media in terms of the equivalent porous medium (EPM) model. *Journal of Contaminant Hydrology* 1:65-76.

Panno, S.V., C.P. Wiebel, P.C. Heigold, and P.C. Reed. 1994. Formation of Regolith-Collapse Sinkholes in Southern Illinois - Interpretation and Identification of Associated Buried Cavities. *Environmental Geology,* 23:214-220.

Park, N.-S., T.N. Blandford, Y.-S. Wu, and P.S. Huyakorn. 1994. CANVAS: A composite analytical-numerical model for viral and solute transport simulation, version 2.0. Herndon, VA: HydroGeoLogic, Inc.

Parker, B.L. 1996. Effects of Molecular Diffusion on the Persistence of Dense Immiscible Organic Liquids in Fractured Porous Media. Ph.D. diss. University of Waterloo, Waterloo, Ontario.

Parker, B.L., J.A. Cherry, and S.W. Chapman. 2004. Field study of TCE diffusion profiles below DNAPL to access aquitard integrity. *Journal of Contaminant Hydrology,* 74: 197-230.

Parker, B.L., R.W. Gillham, and J.A. Cherry. 1994. Diffusive disappearance of immiscible-phase organic liquids in fractured geologic media. *Ground Water,* 32:805-820.

Patriarche, D., E. Ledoux, J.-L. Michelot, R. Simon-Coinçon, and S. Savoye. 2004a. Diffusion as the main process for mass transport in very low water content argillites: 2. Fluid flow and mass transport modeling. *Water Resources Research,* 40(1):W01517:doi: 10.1029/2003WR002700.

Patriarche, D., J.-L. Michelot, E. Ledoux, and S. Savoye. 2004b. Diffusion as the main process for mass transport in very low water content argillites: 1. Chloride as a natural tracer for mass transport—Diffusion coefficient and concentration measurements in interstitial water. *Water Resources Research,* 40(1):W01516:doi:10.1029/2003WR002600.

Peffer, J.R. 1991. Complex Aquifer-Aquitard Relationships at an Appalachian Plateau Site. *Ground Water,* 29:209-217.

Pieper, A.P., J.N. Ryan, R.W. Harvey, G.L. Amy, T.H. Illangasekare, and D.W. Metge. 1997. Transport and Recovery of Bacteriophage PRD1 in a Sand and Gravel Aquifer: Effect of Sewage-Derived Organic Matter. *Environmental Science & Technology,* 31:1163-1170.

Poletika, N.N., W.A. Jury, and M.V. Yates. 1995. Transpot of bromide, simazine, and MS-2 coliphage in a lysimeter containing undisturbed, unsaturated soil. *Water Resources Research,* 31:801-810.

Pollard, D.D., and A. Aydin. 1988. Progress in Understanding Jointing over the Past Century. *Geological Society of America Bulletin,* 100:1181-1204.

Ponzini, G., G. Crosta, and M. Giudici. 1989. The Hydrogeological Role of an Aquitard in Preventing Drinkable Water Well Contamination - a Case-Study. *Environmental Health Perspectives,* 83:77-95.

Potter, P.E., J.B. Maynard, and W.A. Pryor. 1980. *Sedimentology of shale - study guide and reference source.* New York:Springer-Verlag.

Potter, P.E., J.B. Maynard, and P.J. Depetris. 2005. *Mud and Mudstones.* Berlin, Germany: Springer-Verlag.

Powell, K.L., R.G. Taylor, A.A. Cronin, M.H. Barrett, S. Pedley, J. Sellwood, S.A. Trowsdale, and D.N. Lerner. 2003. Microbial contamination of two urban sandstone aquifers in the UK. *Water Research,* 37:339-352.

Powelson, D.K., and C.P. Gerba. 1994. Virus removal from sewage effluents during saturated and unsaturated flow through soil columns. *Water Resources Research,* 28:2175-2181.

Powelson, D.K., J.R. Simpson, and C.P. Gerba. 1991. Effects of Organic-Matter on Virus Transport in Unsaturated Flow. *Applied and Environmental Microbiology,* 57:2192-2196.

PPG Industries. 1995. Phase 2, Draft Site Wide RCRA Facility Investigation Report submitted to U.S. EPA, January 1995. (prepared by IT Corp.).

Pucci, A.A. 1998. Hydrogeochemical processes and facies in confining units of the Atlantic Coastal Plain in New Jersey. *Ground Water,* 36:635-644.

Pucci, A.A. 1999. Sulfate transport in a Coastal Plain confining unit, New Jersey, USA. *Hydrogeology Journal,* 7:251-263.

Pucci, A.A., T.A. Ehlke, and J.P. Owens. 1992. Confining Unit Effects on Water-Quality in the New-Jersey Coastal-Plain. *Ground Water,* 30:415-427.

Quigley, R.M., F. Fernandez, and R.K. Rowe. 1988. Clayey Barrier Assessment for Impoundment of Domestic Waste Leachate (Southern Ontario) Including Clay-Leachate Compatibility by Hydraulic Conductivity Testing. *Canadian Geotechnical Journal,* 25:574-581.

Quigley, R.M., F. Fernandez, E. Yanful, T. Helgason, A. Margaritis, and J.L. Whitby. 1987. Hydraulic Conductivity of Contaminated Natural Clay Directly Below a Domestic Landfill. *Canadian Geotechnical Journal,* 24:377-383.

Quigley, R.M., Q.H.J. Gwyn, O.L. White, R.K. Rowe, J.E. Haynes, and A. Bohdanowicz. 1983. Leda clay from deep boreholes at Hawkesbury, Ontario; Part I, Geology and geotechnique. *Canadian Geotechnical Journal,* 20:288-298.

Quigley, R.M., and R.K. Rowe. 1986. Leachate migration through clay below a domestic waste landfill, Sarnia, Ontario, Canada; chemical interpretation and modelling philosophies. *ASTM Special Technical Publication American Society for Testing and Materials Special Technical Publication* 933:93-103.

Rawling, G.C., L.B. Goodwin, and J.L. Wilson. 2001. Internal architecture, permeability structure, and hydrologic significance of contrasting fault-zone types. *Geology,* 29:43-46.

Rayne, T.W., K.R. Bradbury, and M.A. Muldoon. 2001. Delineation of capture zones for municipal wells in fractured dolomite, Sturgeon Bay, Wisconsin, USA. *Hydrogeology Journal,* 9: 432-450.

Redman, J.A., S.B. Grant, and T.M. Olson. 1997. Filtration of Recombinant Norwalk Virus Particles and Bacteriophage MS2 in Quartz Sand: Importance of Electrostatic Interactions. *Environmental Science & Technology,* 31:3378-3383.

Redman, J.A., S.B. Grant, T.M. Olson, J.M. Adkins, J.L. Jackson, M.S. Castillo, and W.A. Yanko. 1999. Physicochemical Mechanisms Responsible for the Filtration and Mobilization of a Filamentous Bacteriophage in Quartz Sand. *Water Research,* 33:43-52.
Remenda, V.H. 1993. Origin and migration of natural groundwater tracers in thick clay tills of Saskatchewan and the Lake Agassiz clay plain (Canada). PhD. diss. University of Waterloo, Waterloo, Ontario.
Remenda, V.H., J.A. Cherry, and T.W.D. Edwards. 1994. Isotopic composition of old ground water from Lake Agassiz: Implications for late Pleistocene climate. *Science,* 266:1975-1978.
Remenda, V.H., G. van der Kamp, and J.A. Cherry. 1996. Use of vertical profiles of delta O-18 to constrain estimates of hydraulic conductivity in a thick, unfractured aquitard. *Water Resources Research,* 32:2979-2987.
Remenda, V.H. and G. van der Kamp. 1997. Contamination from sand-bentonite seal in monitoring wells installed in aquitards. *Ground Water*, 35: 39-46.
Rhodes, A.J., E.M. Clark, D.S. Knowles, A.M. Goodfellow, and W.L. Donohue. 1950. Prolonged survival of human poliomyelitis virus in experimentally infected river water. *Can J Publ Health,* 41:248-254.
Rijken, P., and M.L. Cooke. 2001. Role of shale thickness on vertical connectivity of fractures: application of crack-bridging theory to the Austin Chalk, Texas. *Tectonophysics,* 337:117-133.
Ritzi, R.W., D.F. Dominic, and K.W. Kausch. 1996. Aquitard distribution in a northern reach of the Miami Valley Aquifer, Ohio, USA; Part 1, Three-dimensional geostatistical evaluation of physical heterogeneity. *Hydrogeology Journal,* 4:12-24.
Roberts, J.R., J.A. Cherry, and F.W. Schwartz. 1982. A Case-Study of a Chemical Spill - Polychlorinated-Biphenyls (PCBs) 1. History, Distribution, and Surface Translocation. *Water Resources Research,* 18:525-534.
Robertson, J.B., and S.C. Edberg. 1997. Natural protection of spring and well drinking water against surface microbial contamination. I. Hydrogeological parameters. *Crit Rev Microbiol,* 23:143-178.
Rodvang, S.J., and W.W. Simpkins. 2001. Agricultural contaminants in Quaternary aquitards; a review of occurrence and fate in North America. *Hydrogeology Journal,* 9:44-59.
Rophe, B., B. Berkowitz, M. Magaritz, and D. Ronen. 1992. Analysis of Subsurface Flow and Formation Anisotropy in a Fractured Aquitard Using Transient Water Level Data. *Water Resources Research,* 28:199-207.
Rowe, R.K., and P. Nadarajah. 1993. Evaluation of the hydraulic conductivity of aquitards. *Canadian Geotechnical Journal,* 30:781-800.
Rowe, R.K., Quigley, R.M. and Booker, J.R., 1995. *Clayey Barrier Systems for Waste Disposal Facilities*, London, U.K.: E & FN Spon (Chapman & Hall).
Rozkowski, A. 1967. The origin of hydrochemical patterns in hummocky moraine. *Canadian Journal of Earth Sciences,* 4:1065-1092.
Rudolph, D.L. 1989. Studies of groundwater flow and solute transport in a clay aquitard near Mexico City. PhD. diss. University of Waterloo, Waterloo, Ontario.
Rudolph, D.L., J.A. Cherry, and R.N. Farvolden. 1991. Groundwater-Flow and Solute Transport in Fractured Lacustrine Clay near Mexico-City. *Water Resources Research,* 27:2187-2201.

Rudolph, D.L., and E.O. Frind. 1991. Hydraulic Response of Highly Compressible Aquitards During Consolidation. *Water Resources Research,* 27:17-30.

Ruland, W.W., J.A. Cherry, and S. Feenstra. 1991. The depth of fractures and active ground-water flow in a clayey till plain in southwestern Ontario. *Ground Water,* 29:405-417.

Runkel, A.C., R.G. Tipping, E.C. Alexander, Jr., J.A. Green, J.H. Mossler, and S.C. Alexander. 2003. *Hydrogeology of the Paleozoic bedrock in southeastern Minnesota.* Report of Investigation 61. Minnesota Geological Survey.

Russell, D.J., and J. Harman. 1985. Fracture Frequency in Mudrocks - an Example from the Queenston Formation of Southern Ontario. *Canadian Geotechnical Journal,* 22:1-5.

Sabourin, L. 1989. Wood-Preservative Migration in Fractured and Unfractured Clay in Winnipeg: Laboratory and Modelling Analysis. Master's thesis, University of Waterloo, Waterloo, Ontario.

Schijven, J.F., and S.M. Hassanizadeh. 2000. Removal of viruses by soil passage: Overview of modeling, processes, and parameters. *Critical Reviews in Environmental Science and Technology,* 30:49-127.

Schmidt, R.T. 1993. Stochastic analysis of contaminant transport through a fractured clay aquitard to an underlying aquifer. Master's thesis. University of Waterloo, Waterloo, Ontario.

Schreiber, M.E., G.R. Moline, and J.M. Bahr. 1999. Using hydrochemical facies to delineate ground water flowpaths in fractured shale. *Ground Water Monitoring and Remediation,* 19:95-109.

Schultz, L.G., H.A. Tourtelot, J.R. Gill, and J.G. Boerngen. 1980. *Composition and properties of the Pierre Shale and equivalent rocks, northern Great Plains region.* U.S. Geological Survey Professional Paper 1064-B.

Schulze-Makuch, D., D.A. Carlson, D.S. Cherkauer, and P. Malik. 1999. Scale dependency of hydraulic conductivity in heterogeneous media. *Ground Water,* 37:904-919.

Schulze-Makuch, D., and D.S. Cherkauer. 1998. Variations in hydraulic conductivity with scale of measurement during aquifer tests in heterogeneous, porous carbonate rocks. *Hydrogeology Journal,* 6:204-215.

Schwartz, F.W., J.A. Cherry, and J.R. Roberts. 1982. A Case-Study of a Chemical Spill - Polychlorinated-Biphenyls (PCBs) .2. Hydrogeological Conditions and Contaminant Migration. *Water Resources Research,* 18:535-545.

Seaber, P.R., J.V. Brahana, and E.F. Hollyday. 1988. Region 20, Appalachian Plateaus and Valley and Ridge. In *The Geology of North America: Hydrogeology.* Edited by W. Back, S. Rosenshein Joseph, and R. Seaber Paul. Boulder, CO: Geological Society of America: 189-200

Sen, M.A., and M.A.W. Abbott. 1991. Hydrogeological Investigation of a Fault in Clay. *Quarterly Journal of Engineering Geology,* 24:413-426.

SEWRPC/WGNHS. 2002. *Groundwater resources of southeastern Wisconsin.* Southeastern Wisconsin Regional Planning Commission Technical Report No. 37.

Shapiro, A.M., and E.A. Greene. 1995. Interpretation of prematurely terminated air-pressurized slug tests. *Ground Water,* 33:539-546.

Shaw, R.J., and M.J. Hendry. 1998. Hydrogeology of a thick clay till and Cretaceous clay sequence, Saskatchewan, Canada. *Canadian Geotechnical Journal,* 35:1041-1052.

Sidle, R.C., B. Nilsson, M. Hansen, and J. Fredericia. 1998. Spatially varying hydraulic and solute transport characteristics of a fractured till determined by field tracer tests, Funen, Denmark. *Water Resources Research,* 34:2515-2527.

Simpkins, W.W. 1989. Genesis and spatial distribution of variability in the lithostratigraphic, geotechnical, hydrogeological and geochemical properties of the Oak Creek Formation in southeastern Wisconsin. Ph.D. diss. University of Wisconsin-Madison, Madison, WI.

Simpkins, W.W., and K.R. Bradbury. 1992. Groundwater-Flow, Velocity, and Age in a Thick, Fine-Grained Till Unit in Southeastern Wisconsin. *Journal of Hydrology,* 132:283-319.

Simpkins, W.W., J.M. Eidem, H.H. Seo, B.L. Johnson, M.F. Helmke, M.R. Weis, T.B. Parkin, M.R. Burkart, and T.B. Moorman. 1996. Confining units are not created equal; Quaternary history and its control on physical and biogeochemical processes. In Proc. of Geological Society of America 28th annual meeting. Boulder, CO: Geological Society of America: 72

Sims, J.D. 1993. Flow and transport through fractured clay till: A laboratory study. Master's thesis. Dept. of Earth Sciences, University of Waterloo, Waterloo, Ontario.

Simunek, J., M. Sejna, and M.T. van Genuchten. 1999. The HYDRUS-2D Software Package for Simulating the Two-dimensional Movement of Water, Heat, and Multiple Solutes in Variably-saturated Media, Version 2.0. Riverside, CA: U.S. Salinity Laboratory, U.S. Agriculture Research Service.

Skilliter, C.C. 2000. A stratigraphic and geochemical investigation of Upper Devonian shale and marl aquitards, west-central Alberta, Canada. Master's thesis. University of Alberta, Edmonton, Alberta.

Sminchak, J.R., D.F. Dominic, and R.W. Ritzi. 1996. Indicator geostatistical analysis of sand interconnections within a till. *Ground Water,* 34:1125-1131.

Smith, L., C. Forster, and J. Evans. 1990. Interaction of fault zones, fluid flow, and heat transfer at the basin scale. In *Hydrogeology of low permeability environments.* Edited by P. Neuman Shlomo, and I. Neretnieks. Hanover, Federal Republic of Germany:Verlag Heinz Heise: 41-67.

Snoeyink, V.L., and D. Jenkins. 1980. *Water Chemistry.* New York: John Wiley & Sons.

Snow, D.T. 1969. Anisotropic permeability of fractured media. *Water Resources Research,* 5:1272-1289.

Sobsey, M.D., P.A. Shields, F.H. Hauchman, R.L. Hazard, and L.W.I. Caton. 1986. Survival and transport of hepatitis A virus in soils, groundwater and wastewater. *Water Science and Technology,* 18:97-106.

Spane, F.A.J., and P.D. Thorne. 1985. Effects of drilling fluid invasion on hydraulic characterization of low permeability basalt horizons: a field evaluation. In *Hydrogeology of Rocks of Low Permeability.* Edited by E. S. Simpson, editor Tuscon, AZ: International Association of Hydrogeologists (USA Committee):30-43.

Starr, R.C., R.W. Gillham, and E.A. Sudicky. 1985. Experimental investigation of solute transport in stratified porous media; 2, The reactive. *Water Resources Research,* 21:1043-1050.

Stimson, J., S. Frape, R. Drimmie, and D. Rudolph. 2001. Isotopic and geochemical evidence of regional-scale anisotropy and interconnectivity of an alluvial fan system, Cochabamba Valley, Bolivia. *Applied Geochemistry,* 16:1097-1114.

Stueber, A.M., and L.M. Walter. 1994. Glacial Recharge and Paleohydrologic Flow Systems in the Illinois Basin - Evidence from Chemistry of Ordovician Carbonate (Galena) Formation Waters. *Geological Society of America Bulletin,* 106:1430-1439.

Stueber, A.M., L.M. Walter, T.J. Huston, and P. Pushkar. 1993. Formation Waters from Mississippian-Pennsylvanian Reservoirs, Illinois Basin, USA - Chemical and Isotopic Constraints on Evolution and Migration. *Geochimica et Cosmochimica Acta,* 57:763-784.

Sudicky, E.A., and E.O. Frind. 1982. Contaminant Transport in Fractured Porous-Media - Analytical Solutions for a System of Parallel Fractures. *Water Resources Research,* 18:1634-1642.

Sudicky, E.A., and R.G. McLaren. 1992. The Laplace transform Galerkin technique for large-scale simulation of mass transport in discretely fractured porous formations. *Water Resources Research,* 28:499-514.

Swanson, S.K. 2002. Hydrogeologic controls on spring flow near Madison, Wisconsin. Mater's thesis. The University of Wisconsin-Madison, Madison, WI.

Tang, D.H., E.O. Frind, and E.A. Sudicky. 1981. Contaminant transport in fractured porous media; analytical solution for a single fracture. *Water Resources Research,* 17:555-564.

Taylor, O.J., and J.W. Hood. 1988. Region 3, Colorado Plateau and Wyoming Basin. In *The Geology of North America: Hydrogeology.* Edited by W. Back, S. Rosenshein Joseph, and R. Seaber Paul. Boulder, CO: Geological Sociey of America: 51-58.

Therrien, R., and E.A. Sudicky. 1996. Three-dimensional analysis of variably-saturated flow and solute transport in discretely-fractured porous media. *Journal of Contaminant Hydrology,* 23:1-44.

Thury, M., and P. Bossart. 1999. The Mont Terri rock laboratory, a new international research project in a Mesozoic shale formation, in Switzerland. *Engineering Geology,* 52:347-359.

Timms, W., and R.I. Acworth. 2002. Origin, lithology and weathering characteristics of Upper Tertiary - Quaternary clay aquitard units on the Lower Murrumbidgee alluvial fan. *Australian Journal of Earth Sciences,* 49:525-537.

Toran, L. 2000. CRAFIT: A computer program for calibrating breakthrough curves of CRAFLUSH, a one-dimensional fracture flow and transport model. *Ground Water*, 38: 430-434.

Tóth, J. 1995. Hydraulic continuity in large sedimentary basins. *Hydrogeology Journal* 3:4-16.

U.S. Environmental Protection Agency. 1987. *Guidelines for Delineation of Wellhead Protection Areas*. Washington D.C.:U.S. Environmental Protection Agency Office of Ground-Water Protection.

U.S. Environmental Protection Agency. 1991. *Wellhead Protection Strategies for Confined-Aquifer Settings*. Washington, D.C.: U.S. Environmental Protection Agency Office of Groundwater and Drinking Water.

United States Environmental Protection Agency. 2000. National Primary Drinking Water Regulations: Ground Water Rule; Proposed Rules. *Federal Register,* 65:30193-30274.

Updegraff, C.D., K.G. Kennedy, A.A. Bakr, C.N. Culver, and J.T. Kam. 1981. Testing low to moderately transmissive zones in basalt rocks. In *Third invitational well-testing symposium; well testing in low permeability environments.* Edited by W. Doe Thomas and J. Schwarz Werner. Berkeley, CA: Lawrence Berkeley Laboratory, University of California: Pages 94-95.

van der Kamp, G. 1992. Evaluating the effects of fractures on solute transport through fractured clayey aquitards. In Proc. of *1992 Conference of the International Association of Hydrogeologists, Canadian National Chapter*. Hamilton, Ont.: International Association of Hydrogeologists.

van der Kamp, G. 2001. Methods for determining the in situ hydraulic conductivity of shallow aquitards - an overview. *Hydrogeology Journal,* 9:5-16.

van der Kamp, G., and H. Maathuis. 1985. Excess hydraulic head in aquitards under solid waste emplacements. In *Hydrogeology of rocks of low permeability*. Edited by P. Neuman Shlomo and S. Simpson Eugene. Association Internationale des Hydrogeologues; Committee of U.S.A. Members.

Van Heerden, J., M.M. Ehlers, W.B. Van Zyl, and W.O.K. Grabow. 2003. Incidence of adenoviruses in raw and treated water. *Water Research,* 37:3704-3708.

Vargas, C., and A. Ortega-Guerrero. 2004. Fracture hydraulic conductivity in the Mexico City clayey aquitard: Field piezometer rising-head tests. *Hydrogeology Journal,* 12:336-344.

Walton, W.C. 1960. *Leaky artesian aquifer conditions in Illinois*. Illinois State Water Survey Report of Investigation 39.

Wang, H.F., and L.R. Myer. 1994. *Underground facility for geoenvironmental and geotechnical research at the SSC site in Texas*. Springfield, VA: U.S. Dept. of Energy Office of Energy Research.

Wassenaar, L.I., and M.J. Hendry. 1999. Improved piezometer construction and sampling techniquies to determine aquitard porewater chemistry. *Ground Water*, 37(4):564-571.

Wassenaar, L.I., and M.J. Hendry. 2000. Mechanisms controlling the distribution and transport of ^{14}C in a clay-rich till aquitard. *Ground Water*, 38(3):343-349.

Watson, B.J., L.H. Motz, and M.D. Annable. 2001. Water budget and vertical conductance for Magnolia Lake. *Journal of Hydrologic Engineering,* 6:208-216.

Weaver, T.R. 1994. Groundwater flow and solute transport in shallow Devonian bedrock formations and overlying Pleistocene unites, Lambston County, southwestern Ontario. PhD. diss. University of Waterloo, Waterloo, Ontario.

Weissmann, G.S., and G.E. Fogg. 1999. Multi-scale alluvial fan heterogeneity modeled with transition probability geostatistics in a sequence stratigraphic framework. *Journal of Hydrology,* 226:48-65.

Weyer, K. U. 1992. *Subsurface contamination by immiscible fluids*. Rotterdam: A.A. Balkema..

Williams, R., C. Teodoru, and D. Mills. 1996. Field methods for evaluating the hydraulic properties of a siltstone aquitard. In *Proceedings of the Tenth National Outdoor Action Conference and Exposition; Aquifer Remediation/ Ground Water Monitoring/ Geophysical Methods*. Dublin, OH: National Water Well Association, National Ground Water Association:. 443-460

Williams, R.E., and R.N. Farvolden. 1967. The influence of joints on the movement of groundwater through glacial till. *Journal of Hydrology,* 5:163-170.

Wills, J., L. Howell, L.D. McKay, B.L. Parker, and A. Walter. 1992. Smithville C.W.M.L. Site: Characterizations of overburden fractures and implications for DNAPL transport. In *Modern Trends in Hydrogeology: Proceedings of the 1992 Conference of the Canadian Chapter, International Association of Hydrogeologists*, Hamilton, Ont.: IAH: 501-515.

Woessner, W.W., P.N. Ball, D.C. DeBorde, and T.L. Troy. 2001. Viral transport in a sand and gravel aquifer under field pumping conditions. *Ground Water,* 39:886-894.

Wolff, R.G. 1970. Field and laboratory determination of the hydraulic diffusivity of a confining bed. *Water Resources Research,* 6:194-203.

Yanful, E.K. and R.M. Quigley 1990. Tritium, oxygen-18 and deuterium diffusion at the Confederation Road landfill site, Sarnia, Ontario, Canada. *Canadian Geotechnical Journal,* 27: 271-275.

Yates, M.V., C.P. Gerba, and L.M. Kelley. 1985. Virus persistence in groundwater. *Appl Environ Microbiol,* 49:778-781.

Yates, M.V., and W.A. Jury. 1995. On the use of virus transport modeling for determining regulatory compliance. *Journal of Environmental Quality,* 24:1051-1055.

Yates, M.V., W.A. Jury, S.R. Yates, D.L. Anderson, L.M. Stark, and P. Sherblom. 2000. *Measurement of virus and indicator survival and transport in the subsurface.* Project #262, Denver, CO: American Water Works Research Foundation.

Yates, M.V., and Y. Ouyang. 1992. Virtus, a Model of Virus Transport in Unsaturated Soils. *Applied and Environmental Microbiology,* 58:1609-1616.

Yates, M.V., and S.R. Yates. 1988. Modeling microbial fate in the subsurface environment. *Critical Reviews in Environmental Control,* 17:307-344.

Yeager, J.G., and R.T. O'Brien. 1979. Enterovirus inactivation in soil. *Appl Environ Microbiol,* 38:694-701.

Young, A., P.F. Low, and A.S. McLatchie. 1964. Permeability studies of argillaceous rocks. *Journal of Geophysical Research*, 69(20):4237-4245.

Young, R.A., and J. Sun. 1995. 3-D Representation of Aquitard Topography using Ground-penetrating Radar. In *Society of Exploration Geophysicists, 65th annual international meeting; technical program, expanded abstracts with authors' biographies*. Tulsa, OK: Society of Exploration Geophysicists: 389-393.

Ziezel, A.J., W.C. Walton, R.T. Sasman, and T.A. Prickett. 1962. *Ground-Water Resources of DuPage County, Illinois*. Illinois State Water Survey and Illinois State Geological Survey.

Zijl, W., and M. Nawalany. 1993. *Natural groundwater flow*. Boca Raton: Lewis Publishers.

Zlotnik V.A., B.R. Zurbuchen, T. Ptak, and G. Teutsch. 2000. Support volume and scale effect in hydraulic conductivity; experimental aspects. In *Theory, modeling, and field investigation in hydrogeology; a special volume in honor of Shlomo P. Neuman's 60th birthday*. Edited by C.L. Winter and D. Zhang. Boulder, CO: Geological Society of America (GSA).

ABBREVIATIONS

AWWA	American Water Works Association
^{14}C	carbon-14
^{18}O	oxygen-18
^{2}H	deuterium
^{37}Cl	chlorine-37
^{3}H	tritium
AwwaRF	Awwa Research Foundation
b	half-aperture width for fractures
b.p.	before present
BTEX	
C	concentration
C_0	concentration at time 0
cm	centimeter
DCM	dichloromethane
Dd	diffusion coefficient
DF	discrete-fracture
DFMD	discrete-fracture matrix diffusion
DNAPL	dense non-aqueous-phase liquid
DOM	dissolved organic matter
Dt	dispersion coefficient
EPM	equivalent porous medium
ft	feet
gal	gallon
h	hydraulic head
in	inch
K	hydraulic conductivity
k	intrinsic permeability
K_h	horizontal hydraulic conductivity
K_v	vertical hydraulic conductivity
ℓ	layer thickness
l	migration distance
L	length, liter
LNAPL	light non-aqueous-phase liquid

m	meter
MCL	maximum contaminant level
msl	mean sea level
n	porosity
NAPL	non-aqueous-phase liquid
nm	nanometer
PAH	polycyclic aromatic hydrocarbon
PCB	polychlorinated biphenyol
PCE	perchlorethylene
q	Darcy flux
s, sec	second
SEWRPC	Southeastern Wisconsin Regional Planning Commission
T	transmissivity
t	time
TCE	trichloroethylene
μm	micron
$\bar{v}$	average linear groundwater velocity
VOC	volatile organic compound
WGNHS	Wisconsin Geological and Natural History Survey
WI	Wisconsin
yr	year